Bhopal's Ecological Gothic

Ecocritical Theory and Practice

Series Editor: Douglas A. Vakoch, METI

Ecocritical Theory and Practice highlights innovative scholarship at the interface of literary/cultural studies and the environment, seeking to foster an ongoing dialogue between academics and environmental activists.

Recent Titles

Bhopal's Ecological Gothic: Disaster, Precarity and the Biopolitical Uncanny by Pramod K. Nayar
Fallen Animals: Art, Religion, Literature edited by Zohar Hadromi-Allouche
Ecological Thought in German Literature and Culture edited by Gabriele Dürbeck, Urte Stobbe, Hubert Zapf, and Evi Zemanek
The Horse in Literature and Film: Uncovering a Transcultural Paradigm by Francisco LaRubia-Prado
Water in Medieval Literature: An Ecocritical Reading by Albrecht Classen
Sustainability and the City: Urban Poetics and Politics edited by Lauren Curtright and Doris Bremm
Ecocultural Ethics: Critical Essays edited by Rayson K. Alex, S. Susan Deborah, Reena Cheruvalath, and Gyan Prakash
Critical Ecofeminism by Greta Gaard
Writing the Earth, Darkly: Globalization, Ecocriticism, and Desire by Isabel Hoving

Bhopal's Ecological Gothic

Disaster, Precarity, and the Biopolitical Uncanny

Pramod K. Nayar

LEXINGTON BOOKS
Lanham • Boulder • New York • London

Published by Lexington Books
An imprint of The Rowman & Littlefield Publishing Group, Inc.
4501 Forbes Boulevard, Suite 200, Lanham, Maryland 20706
www.rowman.com

Unit A, Whitacre Mews, 26-34 Stannary Street, London SE11 4AB

British Library Cataloguing in Publication Information Available

Library of Congress Control Number: 2017956264 | ISBN: 978-1-4985-4045-2 (cloth : alk. paper) | ISBN: 978-1-4985-4046-9 (electronic)

∞™ The paper used in this publication meets the minimum requirements of American National Standard for Information Sciences Permanence of Paper for Printed Library Materials, ANSI/NISO Z39.48-1992.

Printed in the United States of America

Contents

Preface

This book is an attempt to move away from the legal, scientific and biomedical interpretations of the Bhopal Gas Tragedy (2/3 December 1984 to date). Instead it takes as its focus the cultural grammars that represent a disaster. It studies the events around 2/3 December 1984 through their representations in media, using as its primary texts, published eyewitness accounts, documentary films, evidentiary photographs, feature films and protest campaigns with their crowds, slogans and effigies. It examines the tropes and discourses that inform and determine the representation, recall and reprisals of the disaster, and the sites of disastrous effects—from the family to the community, from material bodies to environmental matter.

The aim is not, emphatically, to aestheticize the tragedy—although one could read the cultural texts as adopting an aesthetics of decay, deterioration and death—but rather to illuminate it in different ways so that the rhetoric of tragedy, survivor-crisis and protests may be made visible to students and readers across media, literary and cultural studies. Work on the rhetoric of eco-disaster and social and environmental justice and work in the emerging field of Human Rights and Literary Studies, embodying the ethical turn in literary-cultural studies have begun to appear since the early to mid-2000s. The present book is a modest contribution to such a field.

Bhopal's Ecological Gothic: Disaster, Precarity and the Biopolitical Uncanny is an academic exercise, and does not hope to play the role of either Doctors Without Borders or aid workers. Documenting and examining representations of real-life extreme violence and suffering is a morally complex task, and there is often an unwritten demand that anyone doing so should/could intervene more *materially* (the criticism made of James Nachtwey's war photographs would be a case in point), but it is not one that this academic project is competent to perform.

The book is also written from within a view that modernity's route of "progress" is not always toward freedom, democracy and rights, as post-Enlightenment, liberal, neoconservative and other schools of thought may argue. Rather it aligns its thinking on postcolonial modernity with the writings of Zygmunt Bauman and others working within Human Rights studies who see tyrannical, oppressive and genocidally violent states of being as integral to this presumed "progress." Bhopal instantiates *this* version of modernity.

Acknowledgments

My major debt is to Douglas Vakoch who, as editor of Lexington's Ecocritical Theory and Practice Series, invited me to write a book to follow up on my 2014 essay "From Bhopal to Biometrics." Prof. Vakoch was persistent when I was dithering, and encouraging when I was discouraged. He eventually helped me narrow the focus to the Bhopal tragedy.

Lindsey Porambo at Lexington has been a keenly observant and cooperative editor, for which I owe her many thanks.

This has been one of three hard books in the recent past to write, given its subject matter. Discussing vulnerability, precarity and extreme suffering has been unsettling, and this book, like the rest of the triad (*Human Rights and Literature: Writing Rights* and *The Extreme in Contemporary Culture: States of Vulnerability*) has been made possible with the support of many people.

My parents and parents-in-law who have given up trying to find a proper "explanation" for why I work the way I do remain supportive as always. To them, to Nandini and Pranav ("another book dealing with misery? Really?") I owe much of the necessary life outside the writing.

Materials for a book like this were sourced from multiple locations, and it therefore owes huge debts to those who contributed:

Jayendrina Singha Ray trawled through the JNU library and photocopied materials and couriered them for me.

V. Premlata sourced Raghu Rai's elusive volume, and eyewitness accounts from obscure publishers with grace and cheer.

Satinath Sarangi, in the middle of his courageous work in Bhopal with Sambhavna, found time to send me photographs of the effigies that are integral to the cultural grammar of protest by the victims and survivors. The campaign photographs and advocacy material Satinath and his group has produced unrelentingly for 32 years exemplify Dante's line at the opening of *Inferno*: "Through me is the way to join the lost people." The work Sambhavna, the Bhopal Medical Appeal, the International Campaign for Justice in Bhopal, Chingari do, their warriors battling hopelessness, and to the exemplary courage of the survivors: *this book belongs to them*, even though its trajectory and purpose is a purely academic one.

The International Campaign for Justice for Bhopal granted permission to use their photographs, posters and other materials. To them I send my unquantifiable gratitude, and the hope that this book appeals to all of you.

Kim Fortun sent me her book on *Advocacy after Bhopal* and was enthusiastic about the project.

Padma Prakash put me in touch with Satinath Sarangi, made the crucial phone calls, and asked about the progress of the book. Her comments on a short piece I wrote on Bhopal for *eSocial Sciences* were thought-provoking, so thank you, Padma.

Sudhakar Marathe has unceasingly drawn necessary attention, via Literature and now his photography—which emphasize both beauty and fragility in Nature—to ideas of catastrophe, over the years I have known him. While this book does not go so far as advocacy, it does reflect on vulnerability, and this is partly due to his own work.

To friends and well-wishers who stay concerned, connected, interested and enthusiastic about what I do, my health and general well-being, I can only offer vast gratitude:

Premlata, Ajeet, Debjani, Josy, Haneef, Saraswati, Vaishali

Neelu, a friend for over 25 years now, who ensures a message-a-day (at least)

Molly, "Chechu," for her steady, affectionate good wishes, and her daughter

Ron, my unpaid RA, for the same

Soma Ghosh, who makes it a point to query about work-in-progress and offer assorted home remedies (not for overworking, though)

Ibrahim Ali, a friend since primary school, who remains as committed as he has always been

To Nandana Dutta's insistence on theoretical questioning and her splendid affection, special thanks.

K. Narayana Chandran (K.N.C.) never disappoints me when I ask for bibliographic advice (and including within his lists, texts from Augustine to contemporary Bollywood): to his generous scholarship, I owe more than I can state here.

Anna Kurian, passionately committed since 2002 to all my projects and tenacious in ensuring the rigor of my work and the lucidity of my language, deems this a special book, and this only partly due to being a Bhopali herself. To her keen eye when reading the early drafts, the many discussions and her steadying, affectionate hand, this book's making, like many others, is a witness and a beneficiary.

To KNC and Anna a separate note of thanks for demonstrating that we can *still* talk books and ideas in a university department.

To the anonymous reviewer of the manuscript, the book owes a great deal. For, besides the high praise, the review offered incisive comments and suggestions—all of which benefited the finished product.

* * *

Sections of the arguments in chapter 5 were rehearsed in my "Elementals: The Arts of Bhopal, 1984–2015," *eSocial Sciences*, 29 November 2015, and "From Bhopal to Biometrics: Biological Citizenship in the Age of Globalization," in Scott Slovic, R. Swarnalatha and Vidya Sarveswaran (eds.) *Ecoambiguity, Community and Development: Towards a Politicized Ecocriticism.* Lanham, MD: Lexington Books, 2014, 85–98. A section of the conclusion appeared in the essay "The Child Victim as the Face of Modern Suffering," *The Wire*, 24 September 2016. https://thewire.in/67883/meaning-is-a-physiognomy-or-the-child-victim-as-the-face-of-the-modern-age/. I am grateful to the editors and publishers of *eSocial Sciences*, *The Wire* and Lexington Books for allowing me to use the same.

Introduction

Bhopal, Disaster, Precarity

On 2/3 December 1984 methyl isocyanate (MIC), one of the world's most toxic substances, leaked from the Indian plant of the American pesticide manufacturer, Union Carbide, in the central Indian city of Bhopal. The immediate death toll was computed at about 30,000 and the number of injured, in the vicinity of 500,000 (mainly from genetic mutation-linked illnesses, inherited conditions and delayed onset of diseases due to exposure to the gas), thus making it the world's worst industrial disaster in human memory. The remainder of this story is well-known enough to not require repetition, whether it is the shoddy legal process that gave the victims meagre compensation, the failure of the Indian state, the abdication of Union Carbide of its responsibility, the quiet transfer of the Indian plant (Union Carbide India Limited, UCIL) to Dow which subsequently refused to accept the liability of the disaster and, of course, the continued suffering of several hundred thousand Bhopalis.

Numerous documentaries, some literary texts, advocacy and investigative reports and campaign materials have been produced in the thirty years since the events. Several eyewitness accounts have also been published over the years, some in the form of interviews in periodicals and newspapers. News coverage, especially around the anniversary, has been aplenty. Two volumes of photographs, by Raghu Rai and Francesca Moore, have appeared. Visual materials of effigies burnt annually in Bhopal as part of the continuing campaign for justice, pamphlets and newsletters from forums like the International Campaign for Justice in Bhopal have contributed to the archive around Bhopal. These texts and the archive of suffering they constitute are the subject of the present book.

Literary-cultural studies embodying the "ethical turn" have addressed human rights, democracy, torture and environmentalism in the past few years. Such studies enable us to see the rhetorical and discursive strategies employed in fields like Literature, films or comics that generate cultural models of victimhood, trauma, personhood, the Human, civilization or development. Recent works like Kay Schaffer and Sidonie Smith's *Human Rights and Narrated Lives: The Ethics of Recognition* (2004), Elizabeth Swanson Goldberg and Alexandra Schultheis Moore's *Theoretical Perspectives on Human Rights and Literature* (2012) and Sophia McClennen and Alexandra Schultheis Moore's *The Routledge Companion to Literature and Human Rights* (2016) are key instances of such studies. Studies of apocalyptic and dystopian texts, mostly eco-dystopias in cli-fi (climate change fiction/film) waste and "discard studies" (Susan Signe Morrison 2015) as well as cultural environmental studies (Tippins et al. 2010) are also part of this turn. The emphasis in such studies has been on the textual representations and their hidden or underlying social and political concerns that are immediate and often extreme. Aesthetic and narrative strategies that enable us to see these concerns and the cultural models of victimhood become, then, crucial for such work. This book is situated within such a tradition. It seeks to unpack how the world's worst industrial disaster might be studied in terms of the cultural texts produced about and around it.

The cultural texts around Bhopal's disaster exhibit motifs and themes such as information crypts, biopolitics (of the worst sort), secrecy, repression, landscapes of fear, haunting, abjection and violence. The abject, the uncanny, the monstrous are central to the discourses of disaster, survival and continuing disaster in Bhopal's texts about the events and the aftermath. Bhopal is an ongoing nightmare, a persistent haunting, and much of this revolves around poisoned bodies, genetic codes and the environment. There is slow death and social death in the non-recognition of this violence or the injured bodies, by either the UCIL and its owner-companies (Dow) or the state (India). The thanatopolitics—a politics geared toward the production of death, and the death of many—is frightening, and enduring.

In its iteration of these themes, the cultural texts propose implicitly a *Gothicization* of Bhopal. The texts embody and perform a Gothic, one that may be characterized as a disaster-Gothic or even a survivor-Gothic.

THE BHOPAL GOTHIC

Traditional Gothic texts invoked themes of a return to the past and transgression and often employed an aesthetics of fear. The inventory of motifs above is one that we have come to associate with the Gothic tradition, whether in European or global versions (for a collection of recent essays on the traditions and variant models of the Gothic in literature, science and popular culture, see Spooner 2007, McEvoy 2007 and Punter 2012). Timothy Jones has argued that

> The Gothic is something which is *done* rather than something that simply *is* . . . "Gothic habitus" is a shared way of understanding . . . the Gothic describes a way of writing, a way of reading, a way of thinking about stories, a way of imagining . . . a habitus that orchestrates the generation of various texts and variant readings. . . . (2009: 26–27, emphasis in original)[1]

I use the Gothic as a frame to read Bhopal—as event, as place, as effect—and its precarious culture well aware of this tradition, in order to argue for a *Gothicization* of the disaster, continued disaster and the haunting and terrifying suffering that haunts Bhopal and its residents even thirty years after the events. I take the above formulation of the Gothic "habitus" as a frame of understanding, a way of reading the texts around Bhopal. It is in no way to be taken as a conscious, culturally inflected, deployment of Gothic conventions by the producers of the texts themselves. Fred Botting has argued that:

> Gothic illusions of mortality and the sexed body emphasise bloody corpses, ripped flesh and oozing wounds. Its imagined return to the pulsing reality of the body evokes re-pulsion, a pulsion to the body and of the body, but also away from the body. . . . (2007: 203–4)

The Gothic as a genre deals with the "ruination of human identity," where the human becomes the "ab-human," "continually in danger of becoming not-itself, becoming other." It "stages . . . a species of trauma . . . as the human body collapses and is reshapen" (Hurley 1994: 3–4). Botting and Hurley's formulations enable us to read injury, trauma (itself a term etymologically linked to "wounding") and misshapen human forms in Bhopal's cultural texts. It is the haunting and the precarity in Bhopal's texts, built around bodies, settings (environs, including soil and water bodies), cultural apparatuses when speaking of the prefigurations of the disaster, its progress and its aftermath that invites a Gothicized interpretation. The "Bhopal Gothic"—my shorthand term for the Gothicization of Bhopal—then, is a reading of Bhopal the city and the events occurring therein as approximating to Gothic themes and the aesthetics of decay, death and deterioration. It unpacks the thanatopolitical aesthetics in the representations of injured bodies, the events of the disaster and the enduring, prolonged aftermath.

Gothicization, let me hasten to add, links humans, organic nature such as water, plants and land, and the nonhuman (chemicals, such as MIC or phosgene). Gothicization is about *material* bodies, whether of the human or plant life, and their imbrication with the nonhuman. The process of Gothicization, therefore, is not simply about discursive ghosts or the narrative uncanny.

Gothicization is the iteration of material bodies, their insistence, within networks that involve corporate bodies, the human and the nonhuman. Such a Gothicization of Bhopal centers around and spreads outward, much like the MIC gas on 2/3 Dec. 1984, from the UCIL plant.

It is with industrialization and urbanization, Sara Wasson argues, that a "Gothicization of city space" occurs—she is referring to 19th-century London—with "urban centers . . . transformed by monstrous industry, economic exploitation and the isolation" (2014: 132). Wasson continues:

> A dominant theme of urban Gothic is the incorrigible fragility of modernity: even triumphs of the built environment are precarious, and the collapse of edifices is often a metaphor for the psychological crumbling of those who dwell within them. In urban Gothic, the triumphs of architecture and urban planning are fragile when faced with the destructive forces of entropy and human savagery: as Alexandra Warwick notes, the Gothic city "is also a place of ruins, paradoxically always new but always decaying, a state of death- in-life (Warwick 2009: 251). (132)

The events of 2/3 Dec. 1984 not only demonstrated the collapse of a modern factory's processes but also signaled the collapse of cultural protection and the institutional apparatuses designed to ensure the safety of the inhabitants. The "crumbling" that Wasson rightly gestures to is simultaneously about the collapse of the factory, the social relations, bodies and psyche of a city, that will forever be haunted by that night. As we shall see in the first chapter, the very processes, people and devices meant to work toward the modernization of the city contribute to this "crumbling" effect. However, the principle difference of the Bhopal Gothic from the kind that Wasson describes in the case of 19th-century London is that the decay and collapse of the factory is accompanied not by the psychological crumbling of the inhabitants alone (although this is a key condition of the survivors) but with the *material* decay of bodies and the environment.

Wasson proposes that Gothicized cities carry an "ancient poison" that causes the city's "inhabitants to deteriorate to earlier states of physical and moral development" (132). In the case of Bhopal, there is no ancient poison. In the industrialization of Bhopal, new poisons are injected into the soil and water, into bodies and the social fabric, in the form of UCIL. The deterioration of the inhabitants is dated to the UCIL's founding and subsequent disaster. Yet, thirty years later, the MIC and UCIL's other products do take on the characteristics of an ancient poison because these chemical products do not exit the system, the environment, the city. Bhopal *remains* Gothicized—perhaps a Gothic ruin or remains—because its today remains firmly embedded in its past, specifically the UCIL-driven past.

The Bhopal Gothic, while qualifying as disaster-Gothic or even a survivor-Gothic, is fundamentally an ecological Gothic because the texts continually point to the haunting of Nature—primarily soil, water but also the "natural" bodies of humans and animals—even as they examine the nature of this haunting by unholy liaisons, hybrid bodies, the monstrous, among others, that impact the survivors. These hybrids, hauntings and injuries are, as we shall see, the result of the global asserting/ingratiating itself into and uniting with the local. It is the uneven, disproportionate merger of global corporations with local structures that produces a hybrid monster such as UCIL. That is, Bhopal's texts generate an ecological Gothic when representing the injured bodies, the "slow violence" (Nixon 2011) and the unsustainable world/environs these bodies are embedded in, worlds that are themselves constituted by blurred borders and monstrous affiliations. The Gothic, as the pre-eminent form where we may find the human's encounter with evil or demonic global forces whirl the native human about, where there is transgression and the exploitation of innocence and the persistence of the uncanny, might be repurposed to examine the cultural texts around Bhopal. If the abhuman deformed, sick and decaying Bhopali (Indra Sinha's Animal in *Animal's People*, for instance) is the corporeal instantiation of the corporate-political hybrid that was/is UCIL-Bhopal, the injured and haunted *body politic*, protesting annually for better compensation, is its cultural and collective instantiation.

This monstrous, haunted body politic that is Bhopal, imbricates bodies, machines, chemicals with the state, corporate bodies and transnational capital. Stephen Shapiro has proposed that "capitalist commodification produces an intrinsically Gothic experience" (2008: 30).[2] Although this book does not *explicitly* address the role of capitalist modernity and the Bhopal disaster, it remains a subtext to the sections on "unholy liaisons," the haunting of bodies and nature by a product of industrial processes and transnational greed, the "in-corporated" nature of Bhopal and the mobilization of grief—what I describe as the *Gothicization* of Bhopal through UCIL.

Further, like Indra Sinha's Animal, like the mutant children and the twisted bodies of Bhopal's survivors documented in *The Bhopal Marathon*, the numerous documentaries and protest campaigns, who are "border entities" (Hurley 1994: 24–25)—between animal and human forms—UCIL itself is an abomination, a border entity. As the extended litigation demonstrates, it was neither an American company nor an Indian one, neither a global corporate house (Dow) nor the Indian state's factory. As the analysis of the effigies in chapter 3 demonstrates, the factory-as-abomination is an assertion of the uncanny, because it is both familiar and strange. Thus, the "Bhopal Gothic" as I employ it to read these texts, gestures at abominations, ruptured classifications, destroyed orders and blurred borders wherein categories and binaries of global/local, Indian/foreign, human/animal, savior/destroyer (UCIL as the savior of Indian agriculture) and life/death, break down. The precarious culture of Bhopal and its aftermath—and I take this to mean the prefiguration of disaster, the event of the disaster and the aftermath of the disaster—is the collapse of available frames of knowing. The Gothic is a frame that seeks to explain and examine the anomaly that is Bhopal's spectrality, uncanny hauntings and monstrosity.

However, the injury, the trauma, the haunting of Bhopal's disaster- and survivor-Gothic is best approached, as this book demonstrates, through the frame of precarity.

THE CULTURE OF PRECARITY

The Gothic's dissolving, injured, quasi-formed and hybrid bodies (studied by numerous commentators, from David Punter to Kelly Hurley) in the Bhopal texts take the "form" of abject bodies. Their location in an ecosystem that does not sustain life—in fact, slowly drains them of life—is the subject of all textual representations. The Bhopal texts foreground a social ontology of the Bhopali, one that slides from vulnerability to helplessness, or a state of precariousness. It is an ecosystem and ecology wherein precarity is the order of the Bhopali day. The Bhopal Gothic in its discourse of injurability points less to the bodies of the victim than to the embedding of these bodies in specifically dangerous settings and environs, haunted by a toxic past. That is, the Bhopal Gothic of toxic haunting, specters of destruction, secrecy and repression is cathected on to human bodies *and* the body politic, both of which are, thereby, rendered precarious. Precarious subjects—individuals—constitute the precariat public sphere in Bhopal, even thirty years after the disaster. Bhopal instantiates a *precarious cultural condition* in its texts, from 1984 to the present.

Precariousness is the effect of an exposure to the world, which then inflicts injury (Butler 2004, 2009). Life requires a world that sustains life. As Butler puts it: "there is no life without those conditions that variably sustain life, and those conditions are pervasively social, establishing not the discrete ontology of the person but the interdependency of persons" (2009: 19). Some lives are rendered more precarious, unliveable and their deaths less grievable than that of others (Butler 2004). Bryan Turner underscores the point that all ontology and life is social:

> Who we are is a social process that is always constructed in terms of a particular experience of embodiment. Suffering (a loss of dignity) and pain (a loss of comfort, which we need in order to feel secure and confident) are always intertwined, and so vulnerability is both a physical and spiritual condition. Finally, our experience of the everyday world involves a particular place, a location within which experiences of the body and of our dependency on other humans unfold. (2006: 27)

All humans are vulnerable but only some of them are helpless, argues Adriana Cavarero (2011). Cavarero reminds us that it is "only in the newborn, where the vulnerable and the defenseless are one and the same" (20). But the defenseless are the ones who need active care. In the Bhopal texts this merger of the vulnerable and the defenseless becomes obvious. Helplessness is generated when vulnerable bodies are emplaced in the environment and settings that split open their vulnerability to cause injury. Helplessness is therefore the effect of a set of structural conditions, which in Bhopal's texts is documented as prefigurations of disaster. Exposed to the world, the vulnerable Bhopali was rendered helpless, and generations later, the helplessness haunts the Bhopali body and body politic. This is Bhopal's state of precarious exposure, and its ecological Gothic.

In the cultural texts on the disaster, precarious life is the centerpiece, but it is a precarity that is transmitted across generations. Thus, precariousness is not solely about the events of that night. Rather, it is an ongoing process of *endless* suffering. Adriana Cavarero argues that in torture the "centre of the scene is occupied by a suffering body, a body reduced to a totally available object, or, rather, a thing objectified by the reality of pain, *on which violence is taking its time about doing its work*" (Cavarero 2011: 31, emphasis added). Torture is not just the reduction of the subject to flesh but about the protracted extension of pain and concomitant slow erosion of the body's form, attendant dignity, sense of self and identity—what Cavarero calls "horrorism." Horrorism is the extended precarity of slow violence (such as the poisoning of water and soil), where vulnerability shades into helplessness, as is the case with Bhopal's victims. But this is not all.

Alongside Cavarero's definition of protracted suffering as horrorism, the Bhopal Gothic also terrifies. Posters, effigies and newsletters from the campaigns showcase the twisted, malformed bodies, the sicknesses and extended, intergenerational, i.e., transmissible, suffering—the legacy of 1984. If the protracted suffering is horrorist, the cultural texts inspire and invoke terror through the sheer emphasis on malformed corporeality whose causes and injuries are *simply not visible*. Horror, writes Ruth Anolik, "depends on the visible spectacle, the realized experience, to provide fear." Anolik elaborates:

> Terror . . . is the *frisson* that is provoked by the invisible, by what lurks unseen in the dark. Therefore, I reasoned that the texts would likely provoke horror in response to visible disabilities like bodily deformity, and terror in response to invisible disabilities like sense disabilities and infection. (2010: 8)

Bhopal's visible suffering aside, it is its intergenerational and transmissible contamination that is the source of terror, and which makes its cultural texts, Gothic.

Traditionally, the Gothic texts "mark" spaces, whether urban, suburban or rural, through the presence of haunting bodies of ghosts, revenants, the possessed, or simply the Other (vampires, werewolves, among them). Dale Townshend writes:

> The Victorian Gothic . . . renders the architectural spaces of the earlier Gothic tradition considerably more capacious, daring to display the bodies of its inverts, degenerates, doubles and criminals in their sometimes secretive, sometimes bold perambulations across the urban spaces of the modern city. (Townshend 2014: xliii)

The Bhopal cultural text marks the space of the city as a space where such damaged bodies live. Documentaries, propaganda materials, feature films, photographs, eyewitness accounts all regularly "display" injured bodies. Cameras pan across the injured in their homes and alleys, and interviews are occasions for detailed accounts of the injuries. At the risk of being accused of marketing suffering-porn, these texts with their malformed, incapacitated and injured bodies play a major role in the Gothicization of Bhopal. From the quiet historic city described in almost all eyewitness accounts ("the city of lakes," and such epithets are commonly employed), the cultural texts now represent Bhopal as a Gothic space, haunted by its injured, damaged bodies.

The book moves from anticipations and prefigurations of disaster to the accounts of the events themselves (especially eyewitness accounts) to the aftereffects, aftermath and lingering consequences—what I call "haunting"—of the events. It sees the ecological Gothic as a frame of reading precarious lives and haunted spaces.

This book utilizes print and visual texts, but does not perform any ethnographic study by way of oral interviews (see Suroopa Mukherjee's work in this domain, 2010). It seeks not to aestheticize suffering or disaster, nor does it seek to supplant advocacy discourses and texts. It has a much smaller agenda: to examine how we can "read" the disasters in terms of its *representations* across cultural texts. Thus, the larger aim is to situate disaster and suffering within the remit of literary-cultural studies. It is however informed by the belief that certain aesthetic modes—such as the aesthetics of decay, deterioration and death that is akin to the Gothic aesthetic—are discernible in cultural representations of the disaster and the aftereffects.

The first analytical chapter examines the texts' representations of prefigured disaster. It demonstrates how disintegration, death and the entropic collapse of Bhopal is prefigured in the rhetoric of inherited vulnerability. The final events of 1984 merely *repeat* or *materialize*, with amplified disastrous effects, similar moments from before. It tracks the semiotics of disaster that, read backward from the events of 1984, foretold the disaster years before the events. There is even a descent into irrationality in these texts' recall of the prefigurations, again suggestive of the Gothic.

The second chapter focuses on the narratives of the events themselves. Three principal discourses operate in these representations: the discourse of abjection, wherein the texts document the "making abject" of the city, the plant and the people; the discourse of helplessness of the abject bodies and peoples and finally, the making of deathscapes in and around Bhopal.

Chapter 3 is the first of two chapters that studies the "biopolitical uncanny" in the cultural texts. The uncanny consists of three identifiable dimensions. First, the secrecy and misinformation around the plant, its processes, safety measures and products that constitute the ecosystem of the world of Bhopal. Second, the texts embody a discourse of the monstrous. This monstrous is the hybridization of the local with the global, the monstrous body of UCIL, the Indian state and the world's corporate bodies. Alliances and incorporations constitute the monstrous. Finally, the chapter argues that Bhopal's texts point to the disaster as an autoimmune one.

Chapter 4 continues the examination of the biopolitical uncanny. This focuses on the residual effects of the disaster and the slow violence it unleashes in invisible, encrypted ways, upon the environs, bodies and psyches of the people. The chapter moves from haunted bodies to the state of the immediate environs, the family, and the effects of the disaster in the family. From the haunted families of the texts to the extensive haunted community and the city, the chapter opens out its study of the slow violence and lingering toxicity. Finally, it argues that Bhopal's texts document a haunted body politic.

The concluding chapter, 5, posits a postindustrial Gothic around the Bhopal factory and the haunted realms of the toxic-city. It argues that the UCIL ruin's polychronicity induces nostalgia, perhaps for a secular modernity the industry represents even today. It concludes with a section on the postindustrial uncanny in the representations of UCIL. It also studies the construction of an elegiac sublime around the ruins, and the public insurgencies of grief in the annual protests.

NOTES

1. Jones demands that we see Gothic-as-genre in a historicized sense, so that the newness in the genre may be noted. My purpose here is not to track the cultural production of a Bhopal Gothic, but rather to use the Gothic as a frame and critical apparatus to read the culture of precariousness, injury and haunting in Bhopal's texts. We do not come to these as *Gothic* texts wherein we expect certain themes, tropes and aesthetic forms. Rather, we read from within a Gothic habitus when reading these texts about specific forms of precariousness, the precariat public sphere and precarious bodies which then enables us to understand these themes and tropes. In other words, the Gothic is a way of reading texts dealing with precarious lives.

2. The social transformation effected by a shift to capitalist production, argues Shapiro, results in the making of folk-devils, a (re)turn to supernaturalism and magic and a resurgence in

religious beliefs: "Heightened supernaturalism emerges as a cultural marker of a region's initial appropriation by liberal political economy, but it also reappears with each new turn of the screw, even while the specific form of this discourse alters as the accumulated effects of each capitalist spiral make the prior one's 'Gothic-effects' less viable." (32) Reading texts like *Dracula*, Shapiro argues that "from a culturally materialist vantage, the redefinition of Gothic as a representational response by one core society to fears of losing place to another during times of capitalist phase-transition in the world-system provides a general framework that, in turn, allows for texts to reveal the specific tensions of their own moment." (35)

Chapter One

The Prefiguration of Disaster

The documentary *One Night in Bhopal* announces at an early moment: "In Bhopal, there is an ominous sign of things to come" (2004). The statement, albeit pronounced in retrospect, captures the theme of this chapter: the prefiguration and presentiment of disaster.

In traditional Gothic texts, time is dislocated through repetitions, hauntings, ghostly visitations, the sense of *déjà vu* and *presque vu* (literally, "almost seen"). Writing in retrospect, the Bhopal cultural texts suggest a sense of *presque vu* in the signs around the Union Carbide India Limited (UCIL) plant. This sense of *presque vu* is what the present chapter examines as the prefigurations of disaster in the cultural texts around Bhopal. While not strictly prophetic visions, the narratives seem to inevitably suggest that as early as the 1980s, signs were present in the everyday working of the plant that were ominous, and heralded doom. If the "Gothic has always had to do with disruptions of scale and perspective" (Punter and Byron 2007: 50), then the Bhopal cultural texts are loaded with instances of perspectives determined by ominous signs, and wherein scales of economy, survival and human management are disrupted. It is in these signs of disruption of both scale and perspective that prefigurations of disaster are "read." The degeneration of order—management, safety mechanisms, socio-cultural apparatuses—that marks the Gothic is prefigured in several signs in the Bhopal texts.

The designation 2/3 December 1984, Bhopal, is prefigured in the cultural texts, written after the events, dealing with the disaster in several ways. In the opening paragraph of T. R. Chouhan et al.'s *Bhopal: The Inside Story* (1994) they write, "these factors turned the pesticide factory into a ticking bomb that could explode at any time" (17). In Anees Chishti's journalistic eyewitness narrative he writes, "danger from the Union Carbide factory always lurked in the minds of people of Bhopal" (1986: 33). In the feature film, *Bhopal: A*

Prayer for Rain (2014), the opening scenes show Rajpal Yadav, a poverty-stricken rickshaw-puller in Bhopal, having a minor accident that leaves his vehicle, his primary means of livelihood, broken. A boy makes off with his broken wheel in the confusion. Chasing the boy Yadav comes to a halt right outside a large gate from which uniformed workers stream out. He peers this way and that hoping to catch the urchin who stole the wheel. As the camera focuses on Yadav's face we can see the "Union Carbide" sign, high on a frame/structure inside the gates, in the same frame. Yadav is moving toward the plant as workers leave it, presumably at the end of their shift. His destiny, the narrative suggests, would take him to the plant, and disaster, one day. H. L. Prajapati's *Gas Tragedy: An Eyewitness* (2003), in his account of UCIL's establishment, states blandly, "in 1980 MIC production facility was opened. Guidelines for the design of safety systems were not strictly adhered to" (19). He then lists the faulty machinery, negligence in training workers and processes bypassed and, more significantly, a series of previous gas leaks (19–20). Dominique Lapierre and Javier Moro in *It Was Five Past Midnight in Bhopal* inventory instances of gas leaks in the Virginia (Kanawha Valley) UCIL plant years before the Bhopal incident, thus emphasizing that even the US branch of the company used substandard designs and neglected basic norms of safety, and these resulted in minor incidents and accidents (2001: 53–55).

The above instances invite the theme of this chapter: signs of vulnerability that exist within Bhopal and UCIL well before 2/3 Dec. 1984. While admittedly, these texts were produced after the events, they all concur in proposing that vulnerability and disaster were built into UCIL and Bhopal from a very early period.

This chapter examines the discourse of imminent disaster and the making of vulnerability in the cultural texts of the Bhopal tragedy. As an integral component of the ecological Gothic, the discourse of imminent disaster constructs the place and the people as inheriting particular states of being that predispose them to disaster. The thanatopolitics that we discern in the cultural texts around Bhopal is cast in what I call the Gothicization of Bhopal (or "the Bhopal Gothic"), and an aesthetics of decay, death and continuing disaster in which this book is invested, has its prefiguring moments.

Chris Baldick has argued that the Gothic is marked by a "fearful sense of inheritance in time with a claustrophobic sense of enclosure in space" to produce a "sickening descent into disintegration" (1992: xix). Catherine Spooner writes, "Gothic literature is often defined through its concern with revenant history, whether personal (repressed memories) or collective (ghosts, curses, inheritances, returns from the dead)" (2012: 251). In the Bhopal texts the terrible inheritance is of greed, neglect and incompetence, and the claustrophobia is the effect of a sense of entrapment *around* the UCIL plant, where the plant expands to enclose the world. That is, the plant seems to have enveloped, like the methyl isocyanate (MIC) gas itself, the entire neighborhood, trapping people *within* it until they died. Claustrophobia attacks cause its sufferers to experience choking conditions, as though they cannot breathe in the confined space. This too was the real experience of those who were choked, enveloped in the gas from UCIL. The disintegration is, of course, the collapse of everything in the vicinity of the plant.

Bhopal's inheritance is a "necrospective history" (Baudrillard 1994) of the plant design, in the attitudes toward people and profits and in the very processes that constitute the plant. The disintegration, death and entropic collapse of Bhopal is prefigured in the numerous strands that make up its rhetoric of inherited vulnerability, where the final events of 1984 merely *repeat* or *materialize*, with amplified disastrous effects, similar moments from before. When, for instance, Lapierre and Moro suggest that "America was about to step in where only yesterday the last nawab and his guests had still been hunting tigers and elephants" (71), their oddly orientalist phrasing here serves, perhaps unintentionally, also as a potent metaphor for the legacy UCIL would appropriate: of (American/UCIL) *hunters* and *predators* who killed the innocent inhabitants of the area just as feudal lords and kings/queens did earlier. This is necrospective history with a difference. Where necrospective history was characterized by a "cheap avatar of the original" that constitutes a "form of dilution, of historical entropy" and is made up of "ghost-events" and "phantom events" (Baudrillard 1994: 91), in the case of Bhopal, the avatar represents *amplified* mistakes and *enhanced* destruction, whose roots and sources are firmly in the past.

One further point needs to be made. The signifiers presaging disaster in the Bhopal cultural texts are not extrinsic to the people or the premises, or of alien provenance. As in the case of the traditional Gothic, the cultural texts show how the Bhopal Gothic "exists in part to raise the possibility that all 'abnormalities' we would divorce from ourselves, are a part of ourselves, deeply and pervasively" (Hogle 2002: 12). The portents of disaster are within the system, the "UCIL family." Just as secreted gas tanks and secret information were part of the UCIL Gothic, the abject poverty and the eroded cultural networks were intrinsic to the story.

SIGNIFIERS OF DISASTER

A caparisoned horse, frothing at the mouth, eyes bulging in terror, races out of the gas-filled street (and screen) in *Bhopal: A Prayer for Rain* (2014). In the death-lined streets there is poor visibility and the horse, part of a wedding ritual, running out of the vapours enveloping the town serves as a symbol for what the fog reveals. The primitive terror writ on the animal's visage, and the primordial scenes of death, all come out of the fog of gas. The gas fog is symptomatic of the degeneration of minds, matter and bodies. The poor responses of the technical and administrative mechanisms of both UCIL and the city of Bhopal are symbolized in the spread of a gaseous matter, shapeless, amorphous and penetrating. It is uncontainable matter, and marks the complete collapse of borders: of tanks, walls, pipelines. Matter escapes from UCIL and the result is the erosion of bodily matter of the inhabitants.

There are at least two key signifiers in Bhopal's cultural texts that relate to the prefigurations of disaster, prefigurations that were not seen, or understood, as such, until the night of 2/3 Dec. 1984.

Prefigurations of disaster are represented through a specific semiotics in Bhopal's texts, with the concealing-revealing fog being the first of these signs that gesture at the prefigurations. Every visual cultural text has at least one shot of fumes and fog, symbolic of the noxious gas that killed the city's inhabitants in 1984. Fog—common in northern India in the December season—is the most potent symbol in the semiotics of prefigured disaster because of what it *reveals*, thereby contradicting its very character.

UCIL's cost-cutting, bad managerial practices, the absence of information and the faulty machinery are narrative elements that do *not* add up to a coherent narrative. These are all supposedly disparate, even insignificant, events in the triumphal narrative of UCIL's success. Yet this narrative made of these disparate elements is the hidden plot that reveals itself only in the events of 2/3 Dec. 1984. There exist specific kinds of relations between machinery, policy, people, chemicals and settings, and these relations, hitherto hidden and materially embodied in the architecture of UCIL, become visible on 2/3 Dec. 1984. If one were to link the events of that night with Gothic traditions, it is possible to see the fog of gas as achieving one major effect: it makes visible the hidden narrative of UCIL, the architecture of death and destruction that was concealed in the alignment of these elements.

Writing about the role of fog in urban Gothic fiction, Robert Mighall proposes, "Fog is a supremely sublime element, bringing the essential 'obscurity' that Burke put at the very heart of terror. It also points to the problematic of visibility and knowledge. . . . The fog obscures, but also reveals, the true character of the city" (2012: 56).

Whatever lay hidden in the transactions that constituted UCIL and Bhopal become/s horrifically visible *only* in the poisonous fog that envelops Bhopal. Since these are texts produced after the events, the fog becomes symbolic of both, the knowledge of the true nature of UCIL and the secrecy around it, where the true nature is initially a secret and becomes visible only in the fog. The fog is a symbol of narratives that prefigured the disaster, but were not recognized *as* prefigurations. When things begin to fail in the midst of the spreading fog, we see the hidden narrative emerge, and we understand that the disaster was prefigured in this hidden narrative. As the gas cloud that would become the fog climbs out of the UCIL chimney, we are shown in the films the panic-stricken faces of the control room technicians. Then the camera pans to the noneffective sirens, scrubber, vents, fire hoses and other safety devices. The fog creeping across the city is juxtaposed with the failing devices inside the UCIL plant, and we realise that the fog is the explication of the hidden narrative of flawed designs, human error and human attitudes. Such texts focus on what the events of 1984 revealed, *about* and *from* the pasts of UCIL.

In other words, I am arguing that the fog-as-symbolic-narrative in these texts enables us to see the hidden narrative that ought to have been read as warning signs, or omens, of disaster, but was not. As an aesthetic device that resonates so horrifically with the gas clouds of the real events, the fog is the thanato-aesthetic par excellence. At once classical Gothic and contemporary industrial, the fog is a material reminder and a figural symbol.

The spreading fog is also symbolic of the ugliness behind the civilized exterior of the factory. It renders the civilized, if poverty-stricken, human and animal spaces of the town into a pre-civilized, pre-human, wilderness marked by death, even as, in retrospect, it reveals the greed and deceit narratives underneath the triumphal UCIL story. Just as the fog causes the town's housing to disgorge its inhabitants, it also reveals the secret desires and anxieties behind UCIL.

A second signifier of the prefigured-but-ignored narrative of disaster is a specific ceremony in the cultural texts.

In the films on the Bhopal disaster there is a ritual reenactment of a walk through the ruins of the plant. There are slow pans of the camera over the control room, the E610 tank which exploded, the people who were at the scene, among others. In addition, there are references to earlier accidents in the films and the print texts.

Prefigurations of disaster also involve an almost ceremonial revisitation to the places and events. The ceremony consists of two aspects.

First, the ceremony enacts a specific historical *sequence* of events that *had* occurred in particularly significant ways leading to the disaster: Seepage of water into the tank—the reaction that set in—the pressure building up—first signs of leak—attempts to control—the explosion of MIC through the pipes—disaster envelops Bhopal. Second, in the process of sequencing we also get the *elements* of the disaster: This is the control room that failed that night.

These are safety devices—the pressure gauge, for instance—that failed that night. These are people—Suman Dey, Mukund, Qureishi—who were in the plant that night, and failed to stop the disaster.

Every single cultural text on Bhopal visits the sequence and the elements. This repetitiveness is almost liturgical in the cultural texts. The original sequence of events and elements are assimilated into the present narrative as a liturgical rite, repeated *ad nauseum*. I suggest that it is the liturgical repetition in the present that bestows the *sense* of prefigured disaster on the sequence and the elements.

The liturgical reiteration of the sequence and the elements are of course a later resignification of the past. Yet it is also possible to see this resignification as a set of signs revealed early but not understood except as a vague fear. That is, in the Bhopal texts signs are *seen* but not always understood clearly. Frank Kermode refers to a form of oracular prediction that "at the moment of its announcement may seem trivial or irrelevant, the secret sense declaring itself only after long delay, and in circumstances not originally foreseeable" (1979: 1).

The MIC tanks, the factory, the management halls and the control rooms are sites of secrecy, in sharp contrast to the open (as in lacking privacy) modes of living of the poor in the neighborhood. But with the entry of the fog, those that are inside even the minimally closed off spaces of house and home, spill out, racing for the open spaces hoping for help, salvation and redemption. Just as the gas-fog reveals, by turning inside out, the secrets of UCIL, it everts the city too. Tramping through the desolate ruins today reveals the UCIL's family secrets (again a feature of Gothic texts), that we now recognize.

The narrative offers an *a posteriori* explanation of a set of signs that are *now* read as little-understood signs of imminent degeneration and disaster. Cultural texts on Bhopal are exercises, therefore, in retrospective disambiguation, *removing* ambiguity and clarifying it in the narratives and acceptable signs of the UCIL factory and policies. The plant's safety devices, maintenance systems and policies about which there was both secrecy and ambiguity, it is now clear, were disaster-in-the-making. Reading backwards from the event, the narratives are now able to claim that *signs* of disaster were always embedded in UCIL, but many did not see them, and others did not understand them. When, for instance, Dr. Kumkum Saxena is patronizingly told by her manager in *One Night in Bhopal* that she is being unnecessarily alarmist about MIC and the tanks, we as readers recognize the disambiguation narrative: there were signs of negligence that were ignored and which resulted in the disaster.

Disambiguation also reveals a figural aspect of the narratives around UCIL. The success story of UCIL in the world and in India is disturbed by the *a posteriori* reading of the signs of imminent disaster. The machinery, the tanks and the processes *now* emerge as the figural in the discourses of UCIL's success, the fulfillment of Indian farmers' needs and even safety. The figural, as theorized by Jean-François Lyotard, is "whatever is in the signifying system and yet does not make sense in its rules" (Berthin 2010: 49). The figural "transgresses the order of the visible" (48). As the films and the texts pan over the factory, sometimes as flashback and sometimes the ruins of today, we come to see the machinery and the industrial landscape as a figure that interrupts.

This interruptive figure, especially visible when the camera pans across the former workers, the unit's machinery or the plain landscape, is a many-layered signifier because it is at once a material revenant of the past as well as a living embodiment of what *did not die: the poison.* The *figure* of desolate engineers Chouhan or Dey, the doctor Saxena or the defeated journalist Keswani is a material reminder of the failure of cultural protection: for those tasked with preventing or anticipating a disaster could not do so. If, as the Introduction proposed, disasters are the effect of the failure of cultural protection, Bhopal's cultural text offer us signifiers of this failure.

There is another sense of the disruptive figural as connected to the semiotics of prefigured disaster. More poignantly, the figure of the factory's non-working machinery is one that transgresses the discourse of not only UCIL but Bhopal itself. The figure, seen from the vantage point of the present, was always already an interruption, transgression and "unfitting." The old city of Bhopal with its minarets, tiny streets and the old-worldish ethos has, inserted into its narrative, the figure of UCIL. Representing great progress, employment and opportunity, UCIL in retrospect turns out to have been the figure that did not, ever, fit the visible order of the city.

One further point about these figures. If Bhopal is an ecological Gothic that relies on an aesthetic discourse of haunting, then it is also possible to see Chouhan, Puri, Dey, Saxena and Keswani as *haunted* figures. The ecological Gothic, as already noted, is about the haunting of Bhopal's nature and bodies. But it is also about the haunting of these survivors, haunted by the fact of their failure, of their inability to convince UCIL to act, of their inability to convince the people of Bhopal that UCIL was a ticking bomb. This too is a form of haunting, where the above-mentioned people are figures of the haunting.

Beyond these clear material signifiers, there is a wealth of plot, story and narrative that make up the prefigurations of disaster in the cultural texts.

NEGLIGENCE: A NECROSPECTIVE HISTORY

The cultural texts around Bhopal deploy the theme of negligence that is rooted in the company's past, its necrospective history, and which climaxes in December 1984.

In the very first chapter of their *Bhopal: The Inside Story*, T. R. Chouhan et al. chronicle the company's necrospective history originating in a series of bad management practices, government controls and technological slipups. First, the company's presence in India, dating back to 1969 is given a specific intention: "to make money for the American company, not to serve the interests of the Indian people" (19). Second, they note how UCIL was "allowed majority ownership despite government limitations of foreign investments" (19). Diminishing profits "exacerbated Carbide's already negligent management practices, leading to executive decisions that directly caused the contamination of the MIC storage tank that leaked its contents over Bhopal" (20). On the crucial night, "no evacuation plan was prepared" (20) although, there had been a leak in 1982 "injuring 16 workers and endangering the neighboring community" (21). The chapter concludes with "this history makes clear that the Bhopal disaster was not an 'accident' or an act of sabotage but the result of a series of decisions made by Carbide management in the years leading up to the tragedy" (21). D'Silva notes in his account:

> The next morning, the chief minister of Madhya Pradesh announced that J. Mukund and other members of the plant's senior management—R. B. Choudhury, S. P. Choudhary, K. V. Shetty, and Shakil Ibrahim Qureshi—were taken into custody under section 304 (A) [of the Indian Penal Code], "for causing death by negligence or any rash act." (2007)

In *The Bhopal Reader*, the correspondence between the UCIL's Karmachari Sangh and the UCIL administration reveals this history of negligence. In one letter the Working President of the Sangh writes, "Numerous other accidents have occurred in the Plants due to lack of proper check and replacement of . . . corroded equipment" (Hanna et al. 2005: 12). In response to this DN Chakravarty shifts the debate away from the poor quality of the equipment to worker negligence: "it is a matter of regret that because of unmindful removal of safety mask before decontaminating himself, Mr. Ashraf Mohd. Khan suffered exposure to toxic gas." He concludes with, "we would call upon you to impress all our workmen to follow the safety rules without fail" (Hanna 13–14).

Chouhan et al.'s narrative foregrounds the rampaging entropy in the UCIL plant. Necrospective history such as even Chouhan et al.'s brief chapter captures the spiraling *down* of a dissipated plant, chaos and eventual collapse. But, more importantly for our purposes, the narrative captures a cultural entropy as well. With the use of the term "cultural entropy" I signal the management (both financial and administrative) structures of UCIL and the Indian state that seem to have been quietly eroding with no attempts made to stop the erosion. Chouhan's narrative blames the poor management practices of UCIL but also blames the Indian state, as seen above, for negligence in terms of proper regulations and limitations on what a foreign company can do on Indian soil. That is, the culture of UCIL itself, and that of the Indian state, was severely eroded and thereby generated the negligence. The narrative does not until later (23–24) point to the technical failures: what it points to is a "cultural entropy" of company policies, administrative decision-making and the Indian state's regulatory mechanisms. In other words, the rhetoric of negligence traces the acts of negligence to an abiding cultural entropy around UCIL. Such instances of the cultural entropy around Bhopal's UCIL abound.

Lapierre and Moro write, "this same culture [of ensuring precautions] accommodated some surprising deficiencies" (139). They note that the medical staff hired by UCIL had received no specific training in the pathology of gas-related incidents (139). Lapierre and Moro point to negligence in the field of technical training, but attribute it to the kind of culture UCIL built in the Indian plant. Lalit Shastri in his eyewitness account notes how the Bhopal government allotted land to UCIL: "Overlooking the population pressure and the growing needs of the town, the state government allotted a site to Union Carbide India to establish a potentially hazardous pesticide plant at a stone's throw, barely eight hundred yards, to be precise, from the railway station" (12).

Shastri's account is also firmly within the rhetoric of negligence—in this case the government's negligence over the safety of its own citizens when approving the site for dangerous plants like UCIL.

In Rahul Varma's *Bhopal*, the Chief Minister Jaganlal, previously a lackey of the UCIL chiefs but now occupying a moral high ground after the disaster rants at Warren Anderson:

> It [the disaster] didn't happen in the US because you don't store enormous quantities of MIC at your American factory. It didn't happen there because you have automated monitoring systems: you don't rely on workers' noses to tell you where there's been a leak. It didn't happen there because you have extra safety measures in place—measures you didn't bother to install here. (2004: 56)

This rhetoric of negligence is highlighted with tragic irony when documentaries such as BBC's *One Night in Bhopal* point to the *ostensible* reasons why UCIL came to India: "Union Carbide are a great American corporation and in the '60s and '70s they promoted a dream: to make pesticides that would protect crops and help feed the world's hungry millions." As these words are enunciated the camera, in order to highlight the contrast but also to reinforce the magnitude of the company, lingers over the vent scrubber and other apparatuses all of which, as we know, failed on 2/3 Dec. 1984. It is a "beacon of scientific enterprise," says the voiceover. The imminent tragedy is rendered all the more poignant when we are told that "for the people of Bhopal, many of whom live in poverty, Union Carbide simply promises progress and prosperity." The camera lingers over a lower-middle class section of the city as these words are spoken. Dr. Kumkum Saxena, a doctor with UCIL, is shown anxious about the plant. In a discussion with her boss, she points out that the Americans have "special emergency and evacuation plans and are in constant touch with the local authorities" (she is referring, it emerges, to the American plants of UCIL), "but we've got nothing," she concludes. Her concerns, the voiceover tells us, were "ignored." Its "unwilling[ness] to listen" causes her to quit.

Here the rhetoric of negligence and neglect tracks three major shifts in the UCIL narrative as cultural entropy becomes more and more prominent. First, as the BBC text documents, there is a palpable shift away from the original behavior and attitudes of the company. The company began with the intention of feeding the world's millions by safeguarding crops, as the documentary notes, but then begins to exhibit a frightening callousness and indifference toward the population in the immediate neighborhood of the Bhopal plant. Second, the company drifts away from its emphasis on safety ("Safety Is Everybody's Business"—the sign in UCIL—is highlighted in all documentaries) to negligence and neglect of safety measures. Third, there is a shift in focus from the fertilizer it manufactures—Sevin—and which was meant to protect crops and bring prosperity to the processes and one of the potent constituents: MIC. The shift is tragically ironic when we see the camera linger over a sign that says, "A miraculous pesticide to the rescue of the world's farmers."

Finally, the company itself alters from being a "potential money-spinner to a financial liability." This last is what begins to determine the decisions that then climax in the tragedy of 2/3 December 1984. Themistocles D'Silva documents how "necessary maintenance and repairs had been neglected or discontinued." He then adds, "Employee grievances were not addressed in a timely and professional manner and employee morale was at its lowest." Negligence is therefore not only about the machinery and the technological aspects of UCIL but also the social relations between workers and employers.

The culture of the company changes, the texts suggest, and this is itself a portent of things to come. We see this cultural shift, founded on negligence and greed, critiqued in other, later texts as well. In its response to "The Human Element" campaign of 2006 by Dow Chemicals (which now owns UCIL) the International Campaign for Justice for Bhopal released a poster depicting Bhopalis collecting water from a leaking pipeline (see https://www.bhopal.net/resources/campaign-resources/). The text accompanying the image notes that though UCIL has shut down its chemical residues leak into the local water, poisoning it for over 20,000 people.

Dow promised safe drinking water to everyone on earth in 2013. The promise of a good deed is however belied by its actions, which are determined not by considerations of Corporate Social Responsibility (CSR) but greed. The Bhopal campaign zeroes in on this contradiction of the *rhetoric* of safe water and the *action* of poisoning. If Dow's CSR and its emphasis on care, safety and social justice is the highly publicized verso, the Bhopal campaign reveals its recto: it is precisely care, justice and social concerns that Dow destroys.

The rhetoric of negligence, then, is central to both, the necrospective *history* of UCIL and Bhopal and the imminent disaster in the *future*. Negligence is the process through which this history eventually culminates in the future. In the representation of the ecological Gothic that is Bhopal, the rhetoric of negligence demonstrates how inheritances and legacies determine the horror of 1984. The ecological Gothic is the corruption, over time, of stability, prosperity and good intentions into negligence, callousness and greed. Cultural entropy of the kind UCIL exhibits, suggest the texts, will eventually have adverse impacts on the environment. The collapse of the Bhopal ecosystem therefore stems from UCIL's necrospective history.

It is also possible to read this rhetoric as gesturing at the loss of integrity of the *corporate body* of UCIL. If, as Kelly Hurley (1994) has argued, the Gothic is characterized by the degeneration, disintegration and dissolution of the coherent, bounded human body, the rhetoric of negligence in the Bhopal cultural texts, by pointing to the cultural entropy of UCIL, draws attention to the loss of integrity of the UCIL as a corporate body. The rhetoric of negligence gestures at the loosening or dissolution of ties and commitments to the initial promises and regulations with which UCIL began its existence. It sets out to provide prosperity and employment to the neighborhood, but eventually begins to neglect the very features that were put in place to ensure that the local inhabitants are safe. Negligence here is the abandonment of technical and moral integrity that diminishes UCIL.

Documentaries such as *One Night in Bhopal* show former technicians and engineers of UCIL, Suman Dey and T. R. Chouhan, wandering through the ruins of the *material body* of the company—the Bhopal plant. The ruin, of course, represents the present-day state, but the rhetoric of the films suggests that the today's ruin originated in the neglect and negligence from the days when UCIL was a working unit. The present-day inversion of a factory that highlights abandonment and rejection extends, in fact, the abandonment and rejection of safety procedures, the welfare of the workers and the proper attention to details even when the plant was in working order. The ecological Gothic's (factory) ruins—more about this later, in the reading of postindustrial Bhopal—and disintegrating (human) survivors in Bhopal are contemporary "versions" or manifestations of UCIL's older themes, in other words. Eventually, of course, UCIL and Dow would disown responsibility, thus bringing to a climax a series of negligent acts.

From the disintegrating ties and commitments through the disintegrating factory and finally to the acts of negligence by way of UCIL's abandonment of responsibility for the victims, the pollution and the reparative measures, we see a continuum in the rhetoric of negligence. When we look at this continuum of neglect one can see emerging already the condition that would characterize the events of 2/3 Dec. 1984 and its immediate aftermath: the collapse of cultural protection.

A COMMUNITY OF PREVIVORS

The social world of UCIL comes in for sustained attention in all the Bhopal cultural texts. All Bhopal texts emphasize spatial proximity and cultural distance: of the multinational, billion-dollar UCIL plant with densely populated areas, peopled mainly by the very poor. Documentaries, feature films and commentaries all pan over the poverty-stricken area of hovels, huts and filth as they construct a community of previvors. (I adapt the term from the medical humanities used to describe individuals who, because of inherited genetic anomalies, are *predisposed* to cancer but who have not seen manifestations of the disease, or have survived the predisposition.)

Boris Porfiriev speaks of disaster as "a combination of these external (exogenous) and internal (endogenous) factors that make a social system vulnerable to losing its stability" (1998: 57). The cultural texts around Bhopal depict the prefiguration of disaster through an emphasis on previvor communities rendered *helpless* when they were already *vulnerable*. The vulnerability lies in the poverty and absence of safety nets for the slums around UCIL. These were the people in direct line of the flow of gas on that fateful night. The already vulnerable are rendered helpless due to the collapse of medical health systems and support mechanisms.

There are two key aspects of the rendering of the already vulnerable into the helpless in the cultural texts around Bhopal's previvors: the abject poverty in which they lead their lives contrasted with the multinational corporation's (MNC) greed and the failure of cultural protection.

Poverty and Greed

Numerous shots at the opening of the documentary *One Night in Bhopal* linger over the slums, skinny children and decrepit hutments, drawing attention to the environs of the UCIL plant. "The factory," writes HL Prajapati, was "not ideally located," being "barely two KM away from the railway station" and was surrounded by "slums and colonies" (21). "Our factory," says Warren Anderson's assistant in *Bhopal: A Prayer for Rain*, "is right in the centre of these slums," to which Anderson retorts, "but this is the chief minister's problem . . . slums shouldn't be there."

The cultural texts with their emphasis on the poverty of the people in the vicinity of UCIL draw attention to people and places most likely to be rendered victims because of their always already vulnerable states of being. The ignorance about the materials they handled by those who worked in the

UCIL, and who hailed from the neighboring "slums and colonies," is compounded by their abject state of poverty that forced them into the job.

The cultural texts thus, very early in their narratives, prefigure the disaster as accruing to those who are previvors, with no social or welfare safety nets. Disaster, as commentators have noted, is not the event but the *effects* of the event. In Wolf R. Dombrowsky's formulation: "Disasters do not cause effects. The effects are what we call a disaster" (1998: 14). But these effects are prefigured in the cultural texts as a tension, and conflict, between two different logics in operation: the previvor logic of survival and vulnerability, and the corporate logic of greed and accumulation.

Lapierre and Moro speak of the visit of UCIL's Chairman Bill Sneath to the Bhopal factory in 1976. Sneath, write Lapierre and Moro, "surveyed with pride" the UCIL research unit (101). Yet, this entire edifice, symbolic of so-called UCIL research interests and initiative ignored the locale and location: "What the American president would not see that winter was the jumble of huts, sheds, and hovels that fringed the parade ground like the swelling of a malignant cancer" (102).

Although Lapierre and Moro characterize the hovels and the tenements of the poor as "cancer," it is UCIL that will eventually turn out to be the real malignance and menace. The language of disease and excess used to describe the locality in this account also, however, foregrounds the blindness of the UCIL authorities. This blindness to the location and poverty becomes, then, a prefiguration of the future when UCIL would metamorphose into the malignance haunting, for a long time, the people and the place. The rampaging excess—"cancer"—is not of poverty but of corporatization and its attendant blindness.

Lapierre and Moro also report a conversation with Eduardo Muñoz, the Argentinian agronomist working for UCIL and who was instrumental in setting up the India plant, who is reporting a UCIL Board of Directors meeting:

> "India has a market of three hundred million peasants," the president [of UCIL] immediately declared. "Five hundred million soon," one of the directors promptly advanced. "Don't you worry, Eduardo, we'll sell out five thousand tons and more!" the president insisted with the unanimous approval of all those present. (Lapierre and Moro 85)

Muñoz informs Lapierre and Moro that the Charleston factory was designed to produce 30,000 tons of Sevin a year, and its plant actually stored nearly 120 tons of MIC (87). When consulting with the French and German pesticide manufacturers, he tells them that the Bhopal plant would be storing "twenty-two or twenty-six thousand gallons of MIC." He reports their responses: "All the experts I met went through the roof . . . your engineers are out of their minds" (87).

As Muñoz indicates the purpose of such dangerously high storage levels was just one: numbers, and increasing numbers. In short, greater production and greater profits. H. L. Prajapati declares in no uncertain terms, "It [Bhopal's disaster] was the creation of greed" (23). In this contrast lies the shape of things to come. The plush, well-appointed offices of the American headquarters of UCIL in *Bhopal: A Prayer for Rain* contrasts absurdly with the tea-shop and people in the screen shots above. It is the clash of corporate greed and profiteering of UCIL with the survival greed of the factory's previvor neighborhood.

When cultural texts construct Bhopal city as a community of previvors, they represent the disaster "as a result of the underlying logic of the community," as disaster theorist Claude Gilbert puts it (1998: 6). Gilbert clearly points to disaster as the effect of social vulnerability. In prefigurations of the disaster, the social vulnerability of the Bhopal previvor population is exacerbated by the concomitant greed of the population that owns UCIL: the American MNC. The mutually exclusive logic of the two communities—greed and profiteering of the UCIL community and the survival, welfare-less logic of the Bhopal previvors—come together after 2/3 Dec. 1984 to produce the tragic effects of the events.

There is one more element to the representation of the previvor community in and around UCIL and that is the intracommunity animosity and mutual preying. That is, within the people in the neighborhood, all dependent on UCIL's job and salaries, there exists, as the cultural texts show, severely high competition. Loïc Wacquant has termed this "horizontal abuse and lateral animosity" among even the marginalized (unpaginated). Yadav's rickshaw wheels being stolen, the resentment of neighbours when he gets a job at UCIL are two instances in *Bhopal: A Prayer for Rain*. Even the minimal social cohesion necessary among the poor is lost, Wacquant notes.[1]

Here Wacquant addresses a theme rarely covered, as he admits, "by scholars of urban poverty and the working class because it contravenes their inclination to valorize a disparaged segment of society" (www.loicwacquant.net. 29 Dec. 2016), of the eroding intracommunity networks that would have otherwise offered at least minimal stability. This contributes to the making of the disaster as well because the absence of opposition to UCIL from within the community most likely to be affected by any disaster enables the company to go ahead. The tea-stall occupants—all minimum wage workers in UCIL, and therefore dependent upon the company—who shout at Keswani for poisoning peoples' minds about UCIL in the same film symbolize this lateral animosity.

The Failure of Cultural Protection

The collapse of social relations—a version of the Gothic's degeneration theme—as anticipated in the lack of healthcare or welfare or even life-sustaining environments for the previvors in the vicinity of UCIL, Bhopal, is imminent disaster in that there is a failure of "cultural protection":

> "Cultural protections" may be redefined as realizations of warnings, or, more precisely, of prognoses. . . . On the one hand, they are based on the experiences which have been generated during the evolutionary process of trial and error (Murphy's law represents the highest generalization). On the other hand, they are based on substantiated imaginations, pictures of possibilities, visualizations, and visions. (Dombrowsky 21)

As noted earlier there were instances of mechanical failure and worker ineptitude in the history of UCIL Bhopal. The journalist Keswani wrote a series of articles in which he cautioned the public about the dangers of the badly run UCIL plant. These were all studiously ignored. He described Bhopal as "sitting at the edge of a volcano" (Hanna et al. 14). We see this collapse of cultural protection, and the prefigurations of disaster, in advocate Shahnawaz Khan's letter to UCIL of 4 March 1983: "That waste water that contains poisons flows through a pipe line going through the middle of Jaiprakash Nagar Colony and falls into the sewage canal near the crematorium. This poisonous water poses danger for the residents and cattle" (Hanna et al. 19).

The differences in the structure and operations of the UCIL plant in comparison with the American ones when highlighted also speak of this failure of cultural protections. This comparison comes from UCIL's own internal documents: "The Bhopal facility has been given 1.5 ratings because of expected Naphthol emissions from the solar ponds. . . . The EIA (Environment Impact Assessment) ratings in this case for the [Virginia] Institute, were zero" (Hanna et al. 19–20).

The collapse of cultural protection is the intentional or unintentional erasure of the prognoses of an accident, even though the *history* of similar accidents at UCIL was well known to all. Keswani's imagining of disaster is matched only by the cultural blackout of the imagining. In the films we are told of workers who had been injured or died in the plant, due to various leaks and mechanical errors. Yet, when the spouses and families warn their men against working in the "killer plant" there is utter disregard for the supposedly alarmist opinions. Previous experiences and history do not have any validity, as we see in the case of Rajpal Yadav's desperation to retain his job in UCIL (in *Bhopal: A Prayer for Rain*) despite clear evidence of negligence and callousness in the way the plant is run, and the stories of the workers who died.

Now the Gothic itself is marked by the theme of contamination (Jenkins 2012: 480): "Such contamination robs the individual of consciousness, physical integrity and, ultimately, identity. As social documents, then, Gothic narratives topically address prevailing or sublimated fears of institutions that threaten the essential human" (480).

In the case of Bhopal, the contamination of bodies, integrity, consciousness and identity is the effect of institutional failure—of UCIL and the Indian state. Institutional failure and the collapse of cultural protection render everything about the Bhopal events, then, a Gothic.

The denial of cultural protection by the Indian state, UCIL and the local administration (medical, legal) to those in the immediate vicinity of UCIL in these texts might be read as the "marginality" characteristic of the new urbanisms in the neoliberal era. To return to Loïc Wacquant's thesis about the new urbanisms, we perceive a "territorial stigmatization" (2016: 1082). The neighborhoods of Jaiprakash Nagar and the immediate slums outside of the UCIL compound are perceived and treated differently. Wacquant speaks of "spatial disgrace" (1083) where the supply of essential services is influenced by the public *perception* of these spaces as "urban purgatories" (1083). The cultural texts document the attitudes of the UCIL hierarchies toward the workers who hail from these neighbourhoods. The plant manager's rude behaviour—that officials will behave rudely toward people from these spaces is an argument forwarded by Wacquant (1083–84)—toward Yadav in *Bhopal: A Prayer for Rain* is one such instance.

Territorial stigmatization guarantees the collapse of cultural protection because these spaces are not perceived as requiring any such protection in the eyes of the state or the company. The contrasts between the UCIL guest house or the corridors of the state assembly with the supplicating victims/workers before and after the disaster in these texts point to the unequal symbolic and material values assigned to the two spaces. When the disaster arrives, therefore, it is the urban purgatory around UCIL that is worst hit. Decades later, the environs of this area remain polluted thereby reinforcing the territorial stigmatization: *these* grounds are *contaminated*, *these* people are *contaminated*.

The absence of cultural protection amplifies the social vulnerability of the previvors because there is no attempt to develop prophylactic or therapeutic care for the gases and chemicals that have already killed a few workers. The silence over their deaths means that the events that ought to have, as prognoses, driven innovation and basic safety nets of medical care—cultural responses and protection—*remain* lacking. In the face of a series of accidents such as the gas leak, cultural protections that were put in place to prevent or limit damages in such an eventuality, fail or function inadequately.

Claude Gilbert writes, "if vulnerability has political implications and responsibilities, it directly concerns the ability of political actors, sometimes called decision-making actors, to face critical situations" (6). In the cultural texts, the people in the vicinity of UCIL have no healthcare that any narrative documents. There is also little money to spare for medicines and therapeutic care after the events of 2/3 Dec. 1984. The documentary films as they pan over the slums and the despairing faces of the previvors, also focus on the helplessness of those actors who are tasked with caring for the previvors: the medical authorities in Bhopal. Hamidia hospital is ill-equipped with the necessary oxygen tanks and medicines. More worryingly, as Dr. D. K. Satpathy and Dr. Gandhe at the Hamidia Hospital discover, they had no idea of the constituents of the gas, or possible antidotes. Lalit Shastri writes, "The doctors on duty at the casualty ward had panicked. . . . Finally, after listening to so many people, it became evident that the hospital authorities had never been prepared for such an eventuality . . . the health machinery cracked down completely" (41).

The absence of information will be studied later in this book, but for now we can see how the *social relations* of doctor and patient simply break apart in the face of the events of 2/3 Dec. 1984. Actors and agents such as doctors simply fail their target audience and this generates the effects we call the Bhopal disaster, and yet this failure is not the effect of UCIL's actions alone but the conditions of social vulnerability in which both doctors and patients live. Signs of social vulnerability among the previvor population of Bhopal also constitute a portent of the disaster, because the events of 2/3 Dec. 1984 cause the fragile social relations of agents/actors and the dependents to melt down.

The above arguments clearly gesture at the failure of human actors vis à vis machinery in the production of disastrous effects. When engineers or technicians who are in charge of safety devices fail to operate because of, say inexpertise or indifference, they produce disaster. Or, when the technicians are unable to ensure safety because the devices have been shut down (as was the case with the four key devices in UCIL), then, again, it is the human element at work because the decision to shut down the safety mechanism was taken by *humans* (the managerial level staff of UCIL, as an attempt to cut costs). Lapierre and Moro write about Kamal Pareek, who quit UCIL when they failed to heed his warnings about the neglect of safety mechanisms:

> The most senior were reluctant to suggest solutions that would compromise the safety of their installations. But the pressures were too great, especially when they came from the Danbury head office in the United States. That was how the decision was reached not to change certain parts every six months but only once a year. And to replace any damaged stainless pipes with ordinary steel piping. (201)

The failure of cultural protection, in the form of the failure of human-machine interactions, in other words, generates the disaster of Bhopal. This emphasis prefigures disaster as a social process rather than just a technological/technical one.

The failure of cultural protection in the form of nonfunctioning safety devices and shortage of staff in the control room is a constituent of the prefiguration of disaster if we concede the arguments of Dombrowsky et al. Cultural protection is integral to disaster prevention and the slow and steady erosion of this protection leaves Bhopal open to the disaster of 2/3 Dec. 1984. Disaster is prefigured in the symbolic representations of poverty, distress and abysmal conditions (filmed and documented after the 2/3 Dec. 1984) to draw attention to the social units and contexts in which UCIL operated and failed.

UNCERTAINTY AND THE PUBLIC SECRET OF RISK

Claude Gilbert posits disaster as entering into a state of uncertainty. He elaborates:

> First, disaster is tightly linked to uncertainty that occurs when a danger, whether real or not, threatens a community, and this danger cannot be defined through causes or effects. Second, uncertainty emerging from modern communities is related to their growing complexity. It is the result of the upsetting in the system of meaning, and not the effect of the difficulty of solving problems of accidents or serious dysfunctions. Uncertainty is mainly the product of community organization and not of external factors. Third, we may speak of disaster when actors in modern societies increasingly lose their capacity to define a situation that they see as serious or even worrying through traditional understandings and symbolic parameters. For a community, disaster means the loss of key standpoints in common sense. (9)

In every cultural text on Bhopal there is an emphasis on the complex relations of the neighborhood with the UCIL plant: dependency, hatred, fear and anxiety. Prefigurations of disaster in the form of accidents occurring throughout its history render the social relations of plant workers and families, plant workers and management and the neighborhood unstable so that they find themselves unable to make coherent meaning of what UCIL *is*, or can be. Starting off as a provider of employment, and therefore sustenance, the UCIL plant in these texts metamorphoses into a zone of contested meaning. The disaster of 2/3 Dec. 1984 is not what produces uncertainty: uncertainty is already writ into the system of meanings and perceptual apparatus through which Bhopal views UCIL even *before* the events of that night.

The UCIL plant and the dangers it posed constituted a "public secret." Michael Taussig defines the public secret as what is "generally known but for one reason or another, cannot be easily articulated" (Taussig, cited in Petryna 2002: 69). Adriana Petryna elaborates on the nature of public secrets in the case of the Soviet Union, "Political and economic conditions . . . lead individuals to neglect their bodies in exchange for something: the stability of a household, authority over the 'facts,' survival. Such forms of reasoning are elementary to the life of public secrets" (71).

Even when there is "the public repression of interior states [there are] private ways that individuals drew together knowledge of historical accountability and truth" (Petryna 69). Every cultural text around Bhopal instances cases of injury in the plant's past (Prajapati 19–21):

- The first injury took place on 24 November 1978.
- The second major accident took place in December 1981.
- A month later, on 8 February 1982, 25 workers were severely affected by phosgene leak.
- Again on 22 April 1982, three electricians sustained burn injuries.
- On 5 October 1982, a failure of the mechanical reel in the phosgene pump led to an accident at the MIC plant.
- Again on 14 October 1982, one Shri Sabu Khan was injured.

Anees Chishti writes, "Danger from the Union Carbide factory always lurked in the minds of people of Bhopal" (33). T. R. Chouhan writes, "Around this time, I began to suffer from problems of indigestion, acidity, and a constant choking sensation in my throat. My coworkers too had similar problems. In 1978 there was a big fire at the factory" (27). And (34–37):

- On 26 December 1981, the serious accident occurred.
- Another incident occurred on 9 January 1982.
- On the night shift of 5 October 1982, MIC, methylcarbaryl chloride, chloroform and hydrochloric acid began leaking.

The history of accidents in these texts renders UCIL a source of uncertain meaning. In the documentary and feature films we are shown the poor grateful for employment in the plant, despite clear indications that the working life therein is unsafe. Rajpal Yadav in *Bhopal: A Prayer for Rain* refuses to concede that working in the plant is dangerous. When the journalist Keswani cautions people about the imminent danger he is chased away for trying to discourage people from their sole source of employment, yet uneasy faces fill the screen as they recall the deaths and injuries of their co-workers.

The prefiguration of disaster has its most potent indexical sign in the instances of *bodily injury* in the cultural texts. The horror that would erupt on 2/3 Dec. 1984 would, as now know in retrospect, mark the single largest poisoning of human bodies in world industrial history. In every documentary and eyewitness narrative we are given accounts of the horror at seeing bodies choke, skin burn and people dying. Thus, the risk the plant creates on a day-to-day basis is only perceptible through the recognition of what its products do to human biology. Until this time MIC and Sevin are just chemicals and factory products: their toxic properties are made visible with the first set of injuries. This emphasis on worker recognition of risk only through bodily injury resonates with what at least one commentator has identified as central to horror: that "horror is essentially bio-horror" (Morgan 2002: 3).[2]

The cultural texts thus implicitly document portents of disaster as biohorror in the first injuries and medical crisis in various "bodies." Further, the injured body as indexical sign that things are not quite alright in UCIL's plant anticipates numerous such injured (and dead) in 1984. The injured body prior to and leading up to 1984 is the embodiment of safety and management processes *not* adhered to.

The making of uncertainty, or disruption in meaning-making, may be conjoined with the known histories of accidents and deaths in the UCIL plant. The absence of discussions about these accidents renders the site and the accidents themselves a public secret. It is the public secret everybody knows but nobody spoke about, acknowledged privately but never accorded public visibility or official recognition.

UCIL is at once inside the town and outside of it. In *One Night in Bhopal* and other films our attention is drawn, almost by accident, to the plant, in some corner of the frame, abutting onto the visual field. The location of UCIL, either in the form of the plant's insertion into the visual, the presence of a uniform (of a UCIL worker) or an employee is integral to the theme of the public secret of risk and uncertainty. The town and its topography revolves around UCIL, yet nobody speaks of UCIL as a hazardous topos in *itself*. What is the nature of UCIL as a site within Bhopal?

One way of thinking about the nature of UCIL as a site or topos comes from cultural geography. Minoli Salgado writing about sites of large-scale disappearances or massacres argues:

> Here cartography is determined by ethical rather than geographical coordinates: truth or material facts are suppressed to the point of invisibility and reality is marked by the bodies of the disappeared—a landscape, it could be said, marked by the truth of disappearance and the disappearance of truth. (2013: 210)

Bhopal's landscape, of which UCIL is a part, is one where the unsavory truths of people poisoned by UCIL's gases and chemicals lie buried along with the bodies. These sites are on the map of the land and yet not on it. They constitute the pubic secret of injury and loss, buried under the rhetoric of employment and profits. UCIL in the middle of the thriving slums, the pulsing markets and the throbbing streets is a heterotopic space. These are spaces where the relations between the state and its citizens are inverted, even destroyed. Peter Johnson's interpretation of Foucauldian heterotopias defines them as "sites which are embedded in aspects and stages of our lives and which somehow mirror and at the same time distort, unsettle or invert other spaces" (2013: 790). UCIL is ostensibly about employment and sustenance, but renders itself heterotopic in these narratives when the texts record the deaths and injuries directly traceable to UCIL and its noxious products. In Rahul Varma's play *Bhopal* he highlights the exclusive space that UCIL occupied. UCIL is situated within the "People's Progress Zone," described as "a country within a country. . . . Free of bureaucratic barriers, an area of deregulation" (22). As it turns out, it is in a zone of deregulation because no Indian laws apply to UCIL.

UCIL's meaning for the citizens of Bhopal is unstable, polarized, as noted above, between two positions. It functions as a heterotopic space because the relations between employer and employee are inverted and those between provider and dependent are distorted. It intrudes into all the spaces occupied by the Bhopalis, and alters them irrevocably. That is, the people in the vicinity of UCIL, in the rhetoric of the cultural texts, are transformed in meaning from "people," "dependents" and "employees" into previvors and imminent *victims* through their constant exposure to risk emanating from UCIL. This is the heterotopic inversion attained by UCIL, and one that renders its meaning unstable in the cultural texts. UCIL is set up in the cultural texts as a space and set of processes that recalibrates the biovalue of the Bhopalis.

Thus, even as UCIL itself embodies an uncertain meaning for Bhopal and its inhabitants, the meanings, or biovalues, of the Bhopalis are also left altered, open to risk and injury.

It is also possible to see the existence of a well-ignored public secret also as the failure of cultural protection: in the form of a disavowal of common knowledge. When the neighborhood and the city itself chooses to ignore warning signs, then knowledge becomes repressed, exactly like the dark secrets that UCIL itself chooses to hide away from the workers, the management and the world at large. That is, we can see the collapse of cultural protection in the denial of common knowledge (but not its absence).

A HISTORY OF IRRATIONALITY IN THE SPACE OF REASON

UCIL's investigations, as is well known, claimed human error, or sabotage, as the cause of the disaster: human error by the Indian employees and workers (see chapter 4). The history of errors in the UCIL plant by the workers—most powerfully captured in the death of Ashraf who, in a moment of panic, took off his gas mask and inhaled phosgene—constitutes an additional layer to the prefiguration of disaster.

The structure of UCIL is important here. The American bosses trained a few Indian workers, but as the films show, many of the regular workers were untrained personnel. The lower rung, untrained workers did not recognize the dangers, while the middle and higher level workers (officers, engineers) recognized them, but were powerless to act because they could not persuade the management of UCIL to mend its ways.

Human error and negligence or (alleged) sabotage effectively proposes a descent into chaos and primitive, premodern behavior by workers at a site of ordered, rationalized and mechanized production. Error implies the collapse of supposedly organized safety mechanisms and of the epistemological frames within which UCIL was in operation. This reversal and collapse is a portent.

Errors of judgement and decision-making (of which there were many in UCIL) imply the slide of reason into irrationality in Gothic texts. José B. Monleon explicates this dimension of the Gothic:

> The descent into madness unearths a journey into primitiveness, discovering insanity at the root of "man's" psyche. In this sense, it implies a voyage back in time to a precivilized state. On the other hand, by identifying person with house Poe endows the mansion with a symbolic value: it is reason overpowered by unreason. Hence it also suggests the collapse of the reasonable premises of bourgeois epistemology. In this sense, it alludes to the role of unreason in shaping society's future, thus uncovering a social and political issue. As with the Gothic production, the recall of primitiveness, of "backwardness," announced the *form of things to come*. (2000: 22, emphasis added)

When the UCIL report blames intentional sabotage by workers or human error by them, it effectively exonerates faulty policy by the American headquarters and personnel of UCIL. This also means that the lapse into irrationality—whether born of disgruntled states of mind or plain ignorance—is a feature of the (Indian) worker. By claiming this lapse as the causal factor, UCIL also rehearses an older stereotype of the emotional Indian/oriental who cannot be trusted to follow established, i.e., rationalized, procedure or order.

When the Indian cultural texts document the lapses, they too imply the erosion of reason and rational thinking, but with a specific caveat. The loss of reason of workers is attributed to the working conditions of their lives within UCIL: poverty, survival, poor training, bad management and faulty machinery. But these documents also admit the large causal factor of human irrational behaviour.

These texts are prefigurations of disaster because they dispute the very assumptions on which the UCIL plant, the MIC processes and the safety mechanisms are based: *rational* thinking in the moment of crisis. Knowledge and expertise both fail. Rational processes, supposedly drilled into the workers in training, shut down and instead anger, dissatisfaction and panic—the (primitive) emotions take over. In the documentation of such incidents, then, we see signs of the imminent disaster: the people are *certain* to crack under pressure, if their history is any index. If psychological disintegration is a key feature of the Gothic, then the collapse of appropriate and rational behavior among the workers of UCIL at crisis-time suggests a Gothic turn in the narrative of UCIL in India. The chain of command falls apart, the machinery—administrative, technical—collapses. These are also anticipated in the events before 1984, and hence might be read as prefiguring disaster.

This mix of rational processes and irrational behavior, mechanization and emotionalism is in fact in keeping with the Gothic's essential and persistent "doubleness" and "blurred" polarizations (Hopkins 2005: xi–xii). While UCIL believes its decisions are based on clear economic data, this decision-making sits uneasily with the human element that makes up the plant. Similarly, the emphasis on safety processes that have been tested and rationalized coexist with the tempers, discontent and ignorance of the workers.

The irrationality theme is also then, like the collapse of cultural protection theme, a version of the Gothic's emphasis on degeneration wherein the processes of safety, control and chain of command had degenerated, and so had worker alertness to safety measures. The documentation of allegedly irrational, stupid and ignorant behavior among the Indian workers of UCIL encoded in the cultural texts transforms the history of UCIL into a history of *disjunctions*: between the evolved processes of an industrial unit and inexplicable ignorance and irresponsible behavior of the workers. Just as in sci-fi Gothic where even scientifically constructed creatures and processes are demonized, this history of UCIL's disjunctions prefigures the disaster by pointing to the inexplicable, the irrational and the uncontrollable at the science's core. The horror of 2/3 Dec. 1984 is a manifestation of precisely this unexpected core. That such advanced technology and rationalized mechanical-scientific processes sit on the same continuum as worker ignorance and irrationality is itself a prefiguration—just as in any Gothic the contradictions between rationality and madness is the core of the tale.

The cultural texts then point to the unstated, undocumented repressed—anger and unhappiness among the workers—as the potentially subversive core at the heart of UCIL. The surface gloss, productivity, concerns with safety are undermined by the repressed. Inverting the privilege of rational science, the cultural texts around Bhopal propose that it is this repressed that produces the horror of 2/3 Dec. 1984. I shall return in a later chapter to the theme of the secret, but for now, it is adequate to recognize that the alleged irrational Indian mind/mentality and modes of behavior are at the core of the "modern" UCIL plant.

NOTES

1. Wacquant elaborates, "When the minimal social stability needed to foster mutuality and to buttress solidarity among and across wage earning households evaporates, the poor cannot but prey on the poor; they strive to avoid and distance themselves from their own kind; and they come to openly despise and blame those like them, or those just a notch below them but too close for comfort" (www.loicwacquant.net. 29 Dec. 2016).

2. Elaborating this, Morgan says that "our neglected, marginalized organic life . . . finds symbolic expression in the atavistic, demonic images conjured by macabre literature. Body horror, pain, death, and dismemberment are facts of everyday physical life on the one hand and phantoms of our dreams and imagining at the same time . . . the horror mythos emerges from "the dark backward and abysm of time" and reflects particular aspects of our bio-existential situation; it emanates, that is, from the agonies and exigencies of physical life" (3).

Chapter Two

The Event of Disaster

Having examined the prefiguration of the disaster in the cultural texts around Bhopal, this chapter turns to accounts of the events and the processes themselves.

Three principal discourses operate in the representations of the events of 2/3 Dec. 1984. The first one is the discourse of abjection, wherein the representations focus on the "making abject" of the city, the plant and the people. In the second, these representations document instances of helplessness of the abject bodies and peoples. Finally, the cultural texts also represent the making of deathscapes in and around Bhopal. Bhopal, 2/3 Dec. 1984, may be fruitfully read through these three discourses.

The discourses which, in the first instance, anticipated disaster, now explicate the events that generate the Gothicization of Bhopal, its persistent thanatopolitics that manifests in the slow violence across generations. Just as the Gothic produced its own vocabulary and aesthetic grammar of hauntings, spectres, insanity, the repressed and the uncanny, Bhopal's cultural texts exhibit a similar but not identical vocabulary.

PRELUDE: THE VOCABULARY OF DISASTER

The eyewitnesses and some of the survivors faced with the unimaginable and the unrepresentable when seeking to describe the events take recourse to a vocabulary that delivers to us, examining the events thirty years later, the genealogy of such traumas.

Rajkumar Keswani in one of his cautionary pieces (1982) warning Bhopalis of the accident waiting to happen, had written, two years before the disaster: "Phosgene gas that was used by Hitler in gas chambers, and that is used for the production of methyl isocyanate" (Hanna et al. 15). In a 15 Oct.

1982 letter to Arjun Singh, the then Chief Minister of Madhya Pradesh, Keswani would reiterate this analogy in a different key: "I will not let this city turn in to Hitler's gas chamber" (Hanna et al. 17). Keswani's comparison via the Holocaust and genocide of the so-called alien Jewish race is an instance of terrible irony: it was the foreign chemical "agent" that took apart the local in the case of Bhopal. N. R. Bhandari also employs Keswani's comparison, noting that phosgene was used by Hitler during World War II" (39). He then calls the disaster a "massacre" (18), and later to the Hiroshima bombing: "like the Hibakushas of Hirosshima [*sic*], people marked by Carbide's poisons have to suffer social ostracism" (35). Prajapati comments: "it [UCIL] became the monstrous killer of the 20th century overshadowing the gruesome nuclear holocaust of Heroshmia [sic] and Nagasaki" (2003: 23).

The vocabulary of nuclear disaster and the Holocaust used to describe Bhopal in these accounts adapts and adopts a history itself. We see in the above accounts a shadow archive and an imaginary economy. In "The Body and the Archive" (1999) Allan Sekula refers to the historical reservoir of images that enables us to make sense of any, even a contemporary, image. He refers to this historical reservoir of images as a "shadow archive." The shadow archive structures our way of seeing, in Sekula's theorization, but it also structures the *production* of images and accounts of the disaster. Thus, the Bhopal vocabulary and imagery relies on a shadow archive formed of Hiroshima and the Holocaust.

Prajapati, Keswani, Bhandari and the other eyewitnesses who employ the vocabulary of the Holocaust or nuclear disaster firmly locate Bhopal within a history of human-induced disaster. When this kind of vocabulary is appropriated to speak of Bhopal then, more or less as a consequence, Bhopal becomes an indexical sign too: for the same set of meanings that grow out of the repertoire of images such as Hiroshima and the Holocaust. In other words, these two "improper" (because there is nothing proper about these names) nouns—Hiroshima and the Holocaust—serve as the shadow archive in which to interpret deeper meanings such as greed, hatred, negligence and conquest. Indexical signs depend, as we know from semiotics, the regularity of patterns and the relevance to the present situation. It also depends on the reader's ability to locate a certain property and its correlation with another object or property. When the commentators invoke Hiroshima and the Holocaust they alert us to similar (but not identical) properties, situations and human functioning that produce a Holocaust- or Hiroshima-like Bhopal.

Hiroshima and the Holocaust are examples of cultural trauma as well, and their invocation by eyewitnesses and commentators instantly alerts us to the contexts in which cultural trauma (Alexander et al. 2004, Alexander 2012) occurs: discriminatory practices, social death of some citizens, postcolonial necropolitics, the failure of cultural protection—many of which will be examined here and in the later chapters. Cultural trauma not only produces social death—where citizens are not deemed to be dying or dead because they have not been deemed to be persons in the first instance—but also intergenerational social death, where the effects of the Holocaust, Hiroshima and Bhopal events haunt corporeally, psychologically and culturally, descendants of the survivors or the gas-exposed.

The common vocabulary which, as proposed above, gestures at cultural traumas as different as Hiroshima, the Holocaust and Bhopal, also calls upon us, as Jeffrey Alexander suggests in his work on a social theory of trauma, to identify "new meaningful and causal connections between previously unrelated events, structures, perceptions and actions" (2012: 6). While this is strictly speaking beyond the scope of the present book, the point is well taken: we need to ask, from the common invocation and vocabulary, how and why specific practices and processes—whether racial discrimination or neocolonialism—disempower certain groups, destroy cultural protection and erode the sovereignty of nations. If, as commentators such as Zygmunt Bauman (1989) have argued, events like the Holocaust are an inevitable, even predictable, consequence of modernity, then the employment of a common vocabulary of injustice, suffering and genocide (a term also used in accounts of the Bhopal tragedy) points to the route modernity has taken through the postcolony as well.

This vocabulary, which also generates images of haunting and continued disaster, is given over to a clear materialism, being invested in the human body marked by disease, decay and suffering. These are bodies marked for death, and a slow death at that, in *Animal's People*, the films and the commentaries. The thanatopolitical imaginary constructed out of these images and descriptions foreground the materiality of vulnerable bodies and the materiality of the chemicals that entered the system (human, environmental).[1] The insistence on embodied assembly corporeal activism (see chapter 5) or haunted bodies (chapters 3 and 4) are both, finally, in Bhopal's cultural texts, materialist in their philosophical and political stance. Ultimately, disaster is about material bodies.

THE ABJECT

In many ways, this effigy-burning, a ritual associated with the continuing campaigns for justice, is a metaphor that encapsulates the Bhopal tragedy. The effigy, burning away, opens up, revealing its innards and the burnt, *material*, interior stuffing. The boundaries between inside and outside are burst wide open. The constituents of the "body" lie displayed in their horrific elemental forms. The body totters, fraying under the heat generated from within and without. This is Gothicized Bhopal itself. Gothicization, then, involves the decay and slow death of material bodies, but is closely linked to the decay of cultural protection and the corporate entities that, literally and figuratively, engineered the decay. The burning effigy serves as a visual metaphor for the event-as-abject/abjecting.

"Abject" is both an operation (to make or render something/someone an abject) and a condition (of being abject), and this is the dual sense in which I employ the term throughout this chapter.

The abject, integral to Gothic texts, was famously theorized by Julia Kristeva as (1984) some-thing that we produce in/from our bodies but must throw away so that the purity of the body can be maintained. The abject is what is both inside and yet outside, belongs to the body and yet has to be excluded. It upsets any sense of identity because as Gothic theorist Jerold Hogle puts it, the abject is the "quintessential state of primordial nonidentity . . . the condition of being half inside and half outside the mother at the moment of birth and of being half dead and half alive in that state" (2012). The abject has to be expelled from within. What causes abjection is "what

Effigy from 2001. Courtesy International Campaign for Justice for Bhopal.

disturbs identity, system, order. What does not respect borders, positions, rules. The in-between, the ambiguous, the composite" (Kristeva). It is a "theory of identity crisis and centres on the waste or excessive products of the body" (Bloom 2012: 221). The abject is closely aligned with the grotesque in its emphasis on sheer corporeal *materiality*, and the decay, degradation and collapse of this materiality. The Gothic is, of course, the preeminent genre for the abject and the grotesque. Such a body generates loathing and fear as bodily secretions break through the body's boundaries.[2]

In *social* abjection, as elaborated by Imogen Tyler (2013), Rina Arya (2014) as well as in related work by Zygmunt Bauman on waste (2004), certain sections of the society are "seen to represent a threat, a fact that legitimizes their exclusion from the social fabric . . . are regarded as abject, lowly and despicable and, to return to etymology, are 'cast away'" (Arya 2014: 7). Some people, therefore, are rendered abject *as* social exclusion. Bhopal's operations do not quite instantiate this view of abjection as a politically motivated project of exclusion. However, the Bhopal representations depict abjection as the effect of a variety of processes that waste bodies and spaces, as this section demonstrates.

Bhopal's corporeal *and* social abjection—which includes both, a process and condition of being abjected—might be read at three principal levels in the cultural texts.

Abject Boundary Object

At the heart of the corporeal and social abjection that Gothicizes Bhopal is an object: MIC. MIC is a boundary object in the narratives that trace the events around 2/3 Dec. 1984. A boundary object, as Susan Leigh Star and James R. Griesemer have described it, are "both plastic enough to adapt to local needs and constraints of the several parties employing them, yet robust enough to maintain a common identity across sites" (1989: 393). Further, "the creation and management of boundary objects is key in developing and maintaining coherence across intersecting social worlds" (393). MIC functions as a boundary object because everything about it was contentious, except its toxicity: its appropriate storage mechanisms, its transmission, the antidotes in case of poisoning, etc.

MIC is a boundary object because it is employed within and across technical, economic and social discourses. MIC is the chemical produced within the technological-technical processes of pesticide manufacture. Former UCIL workers like T. R. Chouhan and other commentators detail the technical chemical processes in the course of which MIC is produced (Chouhan 1994: 44–47, Prajapati 17). Thus, MIC is integral to the technical process of UCIL's manufacture of Sevin: MIC *belongs* to the UCIL process.

In terms of economic discourses, the quantity of Sevin produced in UCIL had decreased over the years due to diminishing demand. This meant that the factory was not making enough profits from its manufacturing processes, leading directly to *economy-determined decisions impacting technical processes.* Thus, Chouhan tells us:

> Faulty plant design was worsened by on-site management decisions to bypass approved operating procedures to save time and money . . . these bypasses were often legitimated by rewriting the rules . . . one such example . . . was provisions for shutting down the MIC refrigeration unit when production was not in process. (58)

H. L. Prajapati writes that "production fell and the demand for MIC based pesticides slumped. This led to gradual deterioration of safety conditions and routine maintenance work" (22).

Lapierre and Moro note in their account that by 1982 there was a falling production rate in the Bhopal unit, almost less than half the capacity (193). Consequently, Jagannathan Mukund shut down key safety systems. Mukund's view was that since it was no longer an active factory, these systems were mere indulgence. Lapierre and Moro observe that Mukund completely ignored the sixty tons of MIC still stored in the tanks on UCIL premises (2001: 203).

It is this set of economy-based decisions that affect MIC's qualitative and quantitative shift from essential component to excess. MIC needed to be controlled, as the accounts demonstrate, but the economics of UCIL become instrumental in deciding the *technical* future of MIC as excess and waste. MIC functions as an *abject* boundary object in these circumstances, appropriated by economists and the management—rendered *excess*, *waste* and a *threat* to the very existence of the plant.

On 2/3 Dec. 1984, MIC shifts qualitatively and quantitatively from being a key component of a process to being excess, waste and threat. Forty-two tons of liquid MIC became "a hurricane of gas" (Lapierre and Moro 268). The pressure in the tank climbs steadily (270), and the gas was escaping into the pipes whose "stopcocks controlling access to the safety device were turned off because the factory was not in service" (275). A chemical supposed to be internal to the reaction was now part of an external force and process: dissemination through the air. Lapierre and Moro describe the apparatus emerging from the ground:

> Tank 610's concrete carapace had just shattered, releasing an enormous steel tank which emerged from its sarcophagus like a rocket, stood vertically, toppled, fell, and stood up again. . . . But it had not burst. From a ruptured pipe at ground level a second geyser then erupted, more powerful and fiercer than the first. (277)

Thus, something hidden, internal and essential had become visible, external and excess. Instead of being processed into something else, MIC emerges as MIC itself, one of the world's most potent chemicals unleashed into the open air.

MIC functions in the cultural texts as a hyperobject. I use the term from Tim Morton (2013), who proposed that things that are "massively distributed in time and space relative to humans," such as Black Holes, the biosphere, the Solar System or even an oil field (2013, unpaginated) are hyperobjects. More importantly, Morton argues that even a "very long-lasting product of direct human manufacture, such as Styrofoam or plastic bags" could be hyperobjects. These hyperobjects are viscous (they stick to anything that comes proximate to them), nonlocal, molten (they trouble the notion of fixed space-time constraints), interobjective (they "can be detected in a space that consists of interrelationships between aesthetic properties of objects." MIC functions as a hidden but deeply embedded (viscous) object, occurring between human tissue, chemicals within the human body, the water they consume in Bhopal and the soil. It persists, as we shall see, across generations of bodies, locales and sites.

If the disaster of Bhopal cuts across material and the non-material, the human and the non-human, then the boundary object is embedded in each of these "contexts" that are, in fact, receptacles for the toxin. The Gothicization of Bhopal is the effect of the toxicity of the boundary object, which destroys all cohesion, internal and external, of bodies, relations and the social order.

Social Abjection

The abject boundary object, MIC, is instrumental in generating social abjection in Bhopal. Social abjection is the *being rendered abject* in specific social and political contexts, of being rendered waste, useless and unwanted (Tyler 2013). It is about exclusion and disenfranchisement within structures of power and identity. It is the reduction of humans to *homo sacers*, or bare life (Agamben 1998), with no rights.

MIC as the abject boundary object transforms the organic, dynamic city of Bhopal into a seething cauldron of social abjection.

First, it forces people to leave their homes and families. Crowd scenes of people rushing out of their houses and localities—locales they had lived in for years, with close ties to their neighbors and their contexts—fill all documentary and feature films around Bhopal. Lalit Sastri documents: "I was surprised to find people fleeing in utter panic. They were rushing away in cars, tempos, auto-rickshaws, minibuses and trucks. . . . These scenes occurred throughout the route . . ." (1985: 37).

Second, the very apparatuses meant to ensure the safety of the people—the hospital and the UCIL factory's emergency measures—fail them, leaving them to die from severe injuries. Hospital scenes in these texts represent the survivors and the dying being neglected and left to die—because the doctors do not know the exact constituents of the gas or its antidote, and the UCIL workers do not offer any clear information about either. MIC renders Bhopalis abject in the sense they become, in a matter of moments, unnecessary appendages—indeed a nuisance—to UCIL.

Third, MIC not only forces them out of their homes and sites of belonging, the Bhopalis are treated as extraneous to the UCIL's problems. Cultural texts show the workers holed up inside the UCIL factory (ironically the only safe place in Bhopal that night) expressing relief that the wind was blowing the gas away from the factory (for instance in *Bhopal Express*, 1999). The UCIL management had problems about the employees' anger, as depicted in texts like *Bhopal Express*, and this becomes central to UCIL's perception of and response to the crisis. When Babulal Varma in the film, a UCIL employee himself, runs to the plant and asks the factory management about the antidote, he is roughed up by them as a "trouble maker." Thus, UCIL evaluates the crisis in terms of its employee's history of demands for compensation and greater safety, and responds to Varma's queries for antidotes as though this was one more instance of such "unreasonable" demands from an angry worker. Varma screams at the managerial staff that people are dying in thousands all over the city and they, the UCIL people, are indifferent to the horror. Social abjection is the loss of all forms of power of the Bhopalis as a result of the MIC.

Fourth, and most significantly, social abjection is the *eversion* of Bhopal city itself, the rupture of the *body politic* with the eruption of what had hitherto been repressed: fears, anxieties, and the gas itself. Social abjection is the process through which order and identity in Bhopal are ruptured. It is a *condition* that afflicts the body politic and not just the biological-corporeal body.

UCIL was originally represented, the accounts reveal, as integral to the city of Bhopal, the lives of its residents and the prosperity of the state.

UCIL's initial reception was thunderous in its welcome. "It was considered as one of the prestigious projects of Bhopal" (Prajapati 15). For many migrants, write Lapierre and Moro, Bhopal itself was the "promised land" (Lapierre and Moro 71), and many wished to have at least one family member as an employee of UCIL (Lapierre and Moro 108).

Culturally, too, note Lapierre and Moro, the UCIL activities enabled the integration of Bhopalis into the plant, although the city's residents were unaware of the exact function of the various machinery they saw being put in place (110). UCIL became a part of the space, the culture, and the lives of many Bhopalis. For many, the blue uniform was the source of all *identity*, as the witnesses and former workers declare in Lapierre and Moro. It is this integrated body—of UCIL and Bhopal, American company and Indian workers—that is torn apart leading to abjection.

Anees Chishti, the reporter on the scene, notes that there were anti-American protests in the immediate aftermath of the disaster, perhaps supported by countries and intelligence personnel from the "eastern bloc" (1986: 48). Lapierre and Moro note the fury against Warren Anderson when he arrives in Bhopal soon after the disaster (337–39). Over the years, of course, the anti-Carbide, anti-USA and anti-Anderson rhetoric has amplified.

If the abject is something the body produces but which acquires the status and qualities of waste, it needs to be purged so that the body's integrity is maintained. It is thus at once *of* the body and yet *not of* it. Until 2/3 Dec. 1984, UCIL is proudly *of* the city of Bhopal, it is part of the lives of the inhabitants, the refuge and provider for the families, the economic cornerstone for all of them.

The US company is transformed overnight into the city's enemy, its scourge and oppressor. From a provider of income, profits and identity, it becomes the city's nightmarish capitalist predator. Documents speak of how the UCIL plant would distribute sweets during festivals (Lapierre and Moro 110) and thus offering signs of *belonging* in an organic connection with the city and its people. UCIL does not respect borders or stay alienated as a distant employer. Instead, in the process of bestowing identity upon the people of Bhopal, it also develops an emotive and cultural association with them, breaking down barriers and becoming a part of their lives. It is this very organic linkage and blurred borders, the dependency and the affiliation that then explodes with the events of 2/3 Dec. 1984.

Thus, what was once a matter of pride—the sense of belonging, the rhetoric of the UCIL "family," the cultural and affective membership—is precisely what wastes away that night. Social abjection is the transformation of the organic linkage into *decay*, of the organisms and organs that constitute Bhopal. UCIL's disavowal from the very instant of the disaster—the absence of information (to which I shall turn later in the book), the lack of responsibility distribution and the "escape" of Warren Anderson after his short trip—signifies not just a withdrawal from this organic linkage and cultural membership but a shift in the *quality* of relations itself. Social abjection is the transformation of relations into indifference and irresponsibility, *and*, proceeding from this, the wasting away of the city.

Social abjection is an effect of the disavowal of UCIL's affective links to the city of Bhopal, of its responsibilities and duties, and therefore its transformation into an enemy, a toxic presence at the heart of the body politic of Bhopal. It marks the erosion of the corporate body of the UCIL plant, its gas tanks and technical apparatuses and consequently the *body politic* of Bhopal itself. Social abjection occurs when processes and structures that hold a body politic together turn noxious, erode and corrode and as a result destroy all identities and borders. Where once the blurred borders of UCIL and Bhopal was a source of pride, these very blurred boundaries become a problem when, on 2/3 Dec. 1984, UCIL shuts down its corporate body and disavows all responsibility as Bhopal burns and chokes.

There is one further aspect to the theme of social abjection. UCIL's location in the immediate environs of densely populated poor colonies was, at some point, a matter of concern, as documented in the preceding chapter. However, because of its strategic relation-building exercises, the plant had become a locus of the area. When 2/3 Dec. happens, this locationality, with minimum separation from the neighbourhood, becomes the source of the great disaster. That is, UCIL which functioned as a sort of mother lode for employment and economic survival, becomes the scourge precisely because of its proximity. If the abject is the extrusion and erosion of inside/outside, the location of UCIL as the heart and soul of Qazi Camp and other areas ensures that proximity turns disastrous when tank E610 explodes. Economic, social and cultural proximity is destroyed as the plant turns into a monster.

People dying and the streets lined with dead bodies also become signs of social abjection. The living dead, the survivors (often represented in films, wailing for their dead families), are what erupted *out* of the Bhopal-UCIL relations via MIC. In a sense—and I shall return to this theme—the dead and the living dead are the *repressed* within UCIL's relations with Bhopal: those kept uninformed, in danger and neglect. The social abjection is the eruption of the repressed in the form of dying and dead bodies. Further, given the prefigurations of disaster, social abjection is the eruption of anxieties and fears (voiced by those like Keswani) in Bhopal's psyche.

MIC as abject boundary object, then, disturbs the order, identity and *body politic* of Bhopal's integrated life and relations with UCIL as it energizes a process of social abjection on 2/3 Dec. 1984.

Social abjection is the ejection of hundreds of bodies, who belong to the social body—the body politic of Bhopal—onto the streets, across the river, and spread over the landscape over the first hours of 2/3 Dec. 1984. Central to the Gothic's revelation of the fragility of modernity (Wasson 2014) Bhopal's text make visible this fragility in the form of decay and death, symptomatic of the inner decay and the failure of cultural protection that might have delayed, or prevented, the decay.

Corporeal Abjection

Rahul Varma's play *Bhopal* begins before the events of 2/3 December 1984. It states in no uncertain terms that UCIL had been poisoning its neighborhood for a long time. Izzat's daughter, Zarina, described as possessing "twisted hands . . . heaving chest. . . . Melting skin" (36) is one of the many children born as mutants and monsters due to Carbide's effluents. The first signs of poisoning, Varma suggests, are the abject and abhuman bodies of the people in the vicinity. Varma's is a prehistory of abjection and contamination. The events of 2/3 December 1984 of course produce more abject bodies.

MIC, once inhaled, turns the insides of the body into mush, people drowning in their own fluids due to pulmonary edema. An eyewitness, Lalit Sastri, writes years later: "I was shocked to see the number of people who were lying on the road . . . most seem to have vomited and froth appeared around their mouths" (37). Prajapati documents:

> The team noticed several dead bodies lying scattered on the roads. . . . People in hundreds were standing on roads and feeling a burning sensation in their eyes; coughing badly, and were having breathing problems and vomiting. (32) Patients lay groaning, gasping and vomiting. . . . (37)

Anees Chishti speaks of "fleeing citizens . . . coughing and spitting, some even bleeding from the mouth," "scrambl[ing] toward hospitals" (25). Rahul the cripple, write Lapierre and Moro, "coughed, choked and spewed up blackish clots" (283). In the hospital, "cohorts of blind people stumble[ed] over the bodies of the dying" (315).

One survivor documents her story in "The Ghosts of That Night" in *The Bhopal Marathon*:

> I was two months pregnant and to my horror and shame I felt myself having a miscarriage right there in the middle of the street. . . . My body was covered with blood. Blood was all over me. I couldn't control my bowels and the feces ran down my legs, mixing with the blood. I was covered with my blood and feces and vomit from my children. . . . The road was so thick with dead bodies that we were forced to step on them. The park between the upper and lower lakes was full of bodies lying on the ground. (8–9)

This description of the abject, akin to other similar ones about vomiting, bodily excretions and collapses (also to be found in the film texts) are clearly instances of grotesque realism (to borrow Bakhtin's term) that focuses on the abject human body. The account of the accidental miscarriage, in particular, forces our attention to the border between fertility and death, between life and death. (I shall return, in a later chapter, to the embalmed and preserved fetuses from such miscarriages.)

These instances offer us an entirely different view of modernity itself. In sharp contrast to the pristine machinery, the smooth processes of production and the organized labor attending to the machinery and the processes in UCIL. Thus, the corporeal abject reveals the grotesque aspects of modern machinery, production and processes embodied in the UCIL plant, for, as Sue Vice has proposed, "abjection is the grotesque of modernity" (1997:

175). The victim and witness narratives cited above and the visual narratives of dying people in the feature and documentary films, cast within the grotesque realist aesthetic, evert bodies so that we see the life-threatening conditions in which the abject emerges in Bhopal. Following the work of Sally Minogue and Andrew Palmer on representations of abortions in modern literature, I suggest that we can see these sights and accounts of the corporeal abject in Bhopal's cultural texts through which "the reader is brought up hard against the physical reality of the body" (2006: 115). Modernization and modernity's dark underside *is* the abject body on the street of Bhopal. It Gothicizes the streets, just as it Gothicizes modernity itself.

Later Raghu Rai and other photographers would capture the agonized physiognomies and twisted limbs of the dead. The dissolution of the body—the stuff of Gothic and horror fiction/film—characterizes all representations of Bhopal's events that night.

The dead and the dying, as documented in these texts are no longer human, but abhuman. As Kelly Hurley defines it, the "abhuman subject is a not-quite-human subject, characterized by its morphic variability, continually in danger of becoming not-itself, becoming other" (1994: 4). It is a collapse of the recognizable human form, and is closely linked with the notion of the abject (5). The bodies in Bhopal dissolve into vomit, tears and other fluids, as these productions of the human body emerge outside. Dying is therefore, in Bhopal, the turning inside-out of bodies, the collapse of bodily boundaries.

The violently injured bodies in the immediate aftermath, but also in the continuing discourses of protest, embody the corporeal abject. Witness accounts' horror is the effect of the eyewitness' encounter with the brutalized dead. This horror around the abject feeds into the Gothicization of Bhopal through a particular aesthetics of the abject as well.

In Tom Herron's 1998 account of Northern Ireland's poetry of violence he argues that horror is generated "in the abjected nature of dispersal." Herron proposes that in such poetry revolving around abject bodies, we can perceive three ways of dealing with the brutalized bodies: "displaying the body as spectacle or image, a site upon which we can read political violence; sanctifying the dead body through acts of ritual; conveying a sense of immediacy and intimacy through witness and elegy" (cited in Brewster 2005: 306). Scott Brewster building on Herron's work argues that through the horror born of these depictions of violated bodies, "violence is ascribed a telos, and the brute fact of sudden death and dispersed bodies is shaped into metanarrative" (306). I suggest that when Raghu Rai, Chishti, Prajapati, and others document the dead bodies around UCIL and within Bhopal, they spectacularize the violated bodies. The photographs of ritual burials and cremations—Rai's heart-stopping and iconic photograph is of the burial of a child-victim—sanctify the dead, even if these bodies are anonymized in the process. The anonymity is part of the horror of dispersal. The photographs of the bodies are also, in the year following 1984, built into a narrative about political violence, neglect and the failure of cultural protection. In other words, the spectacle of dead bodies is abjected, in the sense of being evicted but remain a part of the Bhopal imaginary, *and* transformed into a telos for the effigy-burning (itself, as argued, an evisceration of the stuffed form of the effigy), campaigns and protests. It is with the dead that the later haunting of Bhopal, and the campaigns, begins. If the abject is what disturbs identity, order and the system, as Kristeva has argued, then the bodies lying around Bhopal that night have disrupted the order of the city as nothing else, although they also constitute a telos for a future identity of the Gothicized city.

It could be further argued that the corpse, the ultimate abject that "infects" (Kristeva's term) life, in Bhopal produces a second-layer effect beyond the one upon the *individual* eyewitnesses. Kristeva proposes that when we view a corpse, we understand the borders of the living. Nicole A. Diederich argues following Kristeva that "the individual becomes separated within itself due to the attendant horror at the distinction between life and death" brought home by the sight of the corpse (2011: 22). Both Kristeva and Diederich focus on the effect of the abject on individuals. However, in cases like Bhopal where the dead strewn across the landscape impinge upon numerous people, there is a social or collective perception of the abject as well. It could then be argued that a part of the Gothicization of Bhopal has to do with the awareness and experience of horror, by an entire social order, of the distinction between life and death. That is, Bhopal becomes separated within itself as a result of this recognition of the distinction not through the presence of one corpse but an entire town populated by large numbers of corpses. The Gothicization of Bhopal is the division and resultant anxiety within the social order in the face of such massive sights of piled, dispersed corpses. Bhopal's collective and cultural memory, therefore, is linked to the perception of the abject in terms of the corpses.

Abjection in the Air

Corporeal, spatial and social abjection are linked in Bhopal in a manner that recalls Gothic fiction and representations, especially in terms of images of dissolution and the atmosphere that triggers this dissolution. Glennis Byron has argued that the horror in Gothic texts is the "horror prompted by the repeated spectacle of dissolution—the dissolution of the nation, of society, of the human subject itself" (2012: 187). In the events of 2/3 Dec. 1984, as described in the cultural texts, we can discern dissolution as the main theme: the dissolution of the *human form*, of *spatial* boundaries such as inside and outside and finally, the dissolution of the *body politic* when the relations between UCIL and the city break down. In other words, the dissolution of the human form *and* the social order that is Bhopal are mutually constitutive, with both being rendered the abject through the circulation of a certain kind of air that has potent toxins dissolved into it. Spatial, corporeal and social boundaries are infiltrated, contaminated and everted as this composite air spreads, and people are unable to breathe the oxygen they need separate from the MIC that would kill if breathed in. If the abject, as Kristeva defined it, is characterized by the composite, then the atmosphere in Bhopal that night is the most frightening composite condition of all.

I have already argued that the gas/fog symbolizes the hiding/revelatory nature of UCIL and its processes, a symbol of the hidden narrative of flawed designs, human error and human attitudes. I now suggest that the representations of disseminating gas/fog might also be read as a shift in the very instrument of death: air, and what it *comes to mean* for Bhopal that night.

Peter Sloterdijk, examining gas warfare, has argued: the "most fundamental strata of the biological conditions for life into the attack [in gas warfare]: the breather, by continuing his elementary habitus, i.e. the necessity to breathe, becomes at once a victim" (2009: 22–23). The collapsing of social and human boundaries and bodies is engineered by a change in the quality of *air*, so that Bhopalis that night no longer breathe the life-sustaining gases. The emphasis on the circulating gas and the numerous visuals of the foggy, obscuring night in the films are indexical signs of the *presence* of death in the very air. The primary habitus—breathing—has just been turned into the instrument of death. Thus, the cause of death is the very apparatus that keeps humans alive. Lapierre and Moro document this deadly contradiction when writing of the valiant efforts of the four railway employees who waded through the swirling gas, armed with lanterns, seeking to stop a train from entering a Bhopal station swathed in MIC:

> Very swiftly, however, they ran out of breath. In the end the toxic vapours had penetrated their damp cotton compresses. Hyperventilating with the effort, their lungs craved more and more air, the same air that was poisoned with deadly molecules. (298)

Further, the slow dissemination and dissolution of MIC into the Bhopal atmosphere is also an index and an inverted symbol of how integrated UCIL has been into the Bhopali lives. The gas is shown creeping along the streets and entering houses through apertures and openings. It insinuates itself into the home, and thus dissolves the boundaries of inside and outside because air knows no boundaries, unfortunately, as it turns out, for the Bhopalis.

From the economic and social integration that UCIL represented, the inextricable diffusion of MIC into the night air represents the potent *inversion* of this integrative relation. That is, the theme of dissolving human bodies and body politic is captured by the swirling, dissolving fumes of gases which now come to stand in for the *inversion* and *perversion* of the UCIL-

Bhopal link: just as the company was once part of the lifeline of Bhopalis, the gas dissolving into the crucial oxygen in the atmosphere represents the dissolving of that integration into decay and death. Just as oxygen is no longer the life-sustaining gas for Bhopalis but is instead mixed with poisonous MIC, UCIL is no longer the economy-sustaining plant for the city.

The Gothic, as is well established, is about atmosphere, both literal and figurative. In cultural texts around Bhopal, the air/MIC—since it became impossible to distinguish between the two that night—atmosphere comes to be represented as the instrument of decay, death and destruction through the erosion of the boundaries of the human body, space and the relations between UCIL and the city. It means both the air they breathe and the *sense* of fear pervading the Bhopalis. This dual sense is captured in all visual texts where the camera zooms in on the noxious air *and* on the terror-stricken faces repeatedly. The air rises, and streams out of the UCIL plant and swamps the town, forcing people to breathe and die.

The infusion of this air maps UCIL in entirely new ways onto the geography of Bhopal: a mapping determined by the direction and speed of the wind, the heavy nature of MIC, the low temperatures of a Bhopal winter. That is, MIC and the air literally and figuratively reconfigure the atmosphere and geography of the city. The struggling, choking-for-air, bodies on the streets rush about trying to escape the poisoned air also cause the city's alleys and bylanes to resemble in uncanny fashion the labyrinths of Gothic texts as the heroine (usually) runs, seeking escape. The eerie dark with the ghostly, shimmery gas (in the various documentary and feature films) which the survivors hope to outrun is an exact replica of the traditional Gothic atmosphere in these texts.

HELPLESSNESS

Bhopal is the limit case of structural helplessness, wherein shared vulnerability was converted into irreducible, mostly irreversible, helplessness of wounding and injury.

Central to the representation of the Bhopal disaster in its cultural texts is the discourse of helplessness. Vulnerability, Judith Butler proposes, might be a frame to read *all* humanity (2004, 2009). Examining this theme of vulnerability, Adriana Cavarero reading contemporary violence has argued that the vulnerable body is open to two responses: wounding and care. The helpless is one who, instead of care, can only expect *further* wounding.[3]

Another dimension to this representation of the helplessness of the Bhopalis during that night is the dehumanization and concomitant animalization of the human. The moral philosopher Alasdair MacIntyre has argued that "our bodies are animal bodies with the identity and continuities with animal bodies" (1999: 6). However, we see ourselves, the human species, as "other than animal," as "exempt from the hazardous condition of 'mere' animality" (4). MacIntyre terms this a "cultural prejudice" (5), and points out that injury, disability and helplessness aligns the human, in the event of disaster, with animals when they (humans) too are reduced to "mere" bodies, open to injury.[4]

The panic-stricken horse, the roadsides with dead cattle, the suffering and gasping smaller animals occupy the same representational space as the humans who, in these texts, suffer the same agonizing deaths, the sudden loss of autonomy over their bodies and the helplessness that seeks aid. Lapierre and Moro report: "A white horse, bridled and saddled as if for some celebration. . . . With its eyes bloodshot, its nostrils steaming with burning vapors and its mouth foaming with greenish vomit, the animal bolted away . . ." (287). The Bhopal texts document the collapse of the sustaining and sustainable world of the city that results in the shift from vulnerability to helplessness.

First, as already noted, as the events unfold in Bhopal and the immediate vicinity of UCIL, the "fundamental strat[um] of the biological conditions for life" (Sloterdijk, quoted above) becomes the killer: air. Images of animals frothing at the mouth, eyes bulging in panic and humans in more or less the *same* conditions of distress and agony fill the representations of disaster. The vulnerability of humans to any kind of alteration in the biological conditions for life—breathing—is now rendered into helplessness: they cannot breathe, but they cannot *not* breathe either. What the cultural texts show is: death does not occur due to the gas but due to the human inability to *not breathe in* the gas. In other words, the helplessness is the effect of the dependency of the human body upon its neighborhood, upon environments and the quality of air. The Bhopali, as the texts show, is helpless because he is effectively *passive* and cannot *escape* breathing. As long as s/he breathes the wounding will continue and will progress. The Bhopali is "absolutely exposed" (Cavarero's term) to the injury by the very nature of the human body: it breathes.

Second, vulnerability becomes helplessness because, in the face of wounding when the injured seeks care and aid, s/he cannot find it *in the world*. That is, the vulnerable human who redefines her/his relations with the world during the disaster as only dependency, expects care and comfort. In Bhopal that night, the texts show, there was no care forthcoming. This means, the relation of the human with the world—constitutive dependency—had collapsed completely. The hospital scenes in all the movies highlight this. The choking and gasping victims rush to the Hamidia hospital, and plead with the doctors for relief and treatment. The doctors plead helplessness themselves since they are unaware of the *nature* of the toxin and so cannot administer antidotes, if any. That night, as the texts show, the world of Bhopal was poisonous and deadly, and no aid or succor was coming from it.

Third, vulnerability in these texts when it evolves into full-fledged helplessness, aligns the human with the animal. HL Prajapati juxtaposes two photographs: of a mass of human bodies and of animals on the same page with the latter carrying the caption, "animals were equal sufferers" (36). Numerous photographs of animal bodies in the disaster are available (Prajapati 37, 43, 44). Anees Chishti writes: "in Chhola we saw a large number of cattle and other animals lying dead, not far from human corpses" (26).

If, as Alisdaire MacIntyre has argued (2002), we are our bodies, then this corporeal vulnerability and openness to injury is what we share with animals as well. In Bhopal, that night, an amplification of the corporeal through pain and suffering was common to all humans, and to humans and animals. Dirtying oneself, unconsciously and uncontrollably, aligns the humans that night with the animals.

Fourth, it is important to note how, if vulnerability presupposes ethical responses and compassion, Bhopal exemplifies the difficulties in even thinking of such responses. That is, vulnerability offers a departure point for social relations when we *recognize shared potential for injury*. Judith Butler writes:

> Each of us is constituted politically in part by virtue of the social vulnerability of our bodies—as a site of desire and physical vulnerability, as a site of a publicity at once assertive and exposed. Loss and vulnerability seem to follow from our being socially constituted bodies, attached to others, at risk of losing those attachments, exposed to others, at risk of violence by virtue of that exposure. (2004: 20)

Ann Murphy (2011) responding to both Judith Butler and Adriana Cavarero has argued that the ontological fact of vulnerability can generate ethical obligations.[5] Thus social relations can be reconstituted ethically when we recognize mutual dependency and vulnerability. In Bhopal this ethical response becomes the single most difficult task because the events render everybody helpless and not just vulnerable.

For instance, as noted above, physicians tasked by the Hippocratic oath to offer care and alleviatory mechanisms in the face of suffering, find themselves unable to do so. More worryingly for the physicians at Hamidia, they were uncertain of the efficacy of their treatment: whether these would cure or aggravate the conditions was a moot point. This is where vulnerability conjoins with helplessness in a manner that spells complete disaster. The actions of doctors were acts of *helplessness* in the face of the disaster, with actions leading to uncontrollable and unpredictable consequences in the face of the ignorance around the gas. The ethical response to the injury was both impossible and unpredictable (if performed). Hannah Arendt writes in *The Human Condition*: "While the various limitations and boundaries we find in every body politic may offer some protection against the inherent boundlessness of action, they are altogether helpless to offset its second outstanding character: its inherent unpredictability" (1998: 190).

The helplessness that arises within and as the consequences of one's actions, as Arendt points out, is one of the most troubling features of Bhopal's night of horror. While hospitals, the police and the welfare mechanisms are bounded in their actions and spheres of influence, the helplessness of Bhopal stems from the unpredictability of the events *of* and *in* these actions themselves. Within the UCIL plant, using water, shutting down valves, sounding the siren are actions whose consequences spiral out of control and beyond the spheres of influence of the engineers in the plant. Thus the ethical responses to injury and vulnerability are themselves acts of helplessness.

Fifth, doctors and those seeking to help and alleviate the misery of the victims that night encounter and experience helplessness in a very specific way here in these texts. Citizens, framed as those who have the right to protection, safety and medical care in the city of Bhopal, are suddenly framed differently. This difference is not only between the citizen as citizen and citizen as *biological* victim, but also between the very frames of perception.

Writing about precarious and grievable lives, Judith Butler has argued that the "recognizability" of persons *as* persons, involves having norms and frames within which "life" is defined. Any life that fits these norms is defined as *life*. This recognizability itself might be read along two lines, "apprehension" and "intelligibility." Apprehension "impl[ies] marking, registering, acknowledging without full cognition. If it is a form of knowing, it is bound up with sensing and perceiving, but in ways that are not always—or not yet—conceptual forms of knowledge. . . . A mode of knowing that is not yet recognition" (2009: 5–6). Intelligibility is "understood as the general historical schema or schemas that establish domains of the knowable" (6). What is "apprehended as 'living' . . . is not always recognized as a life" (8). Frames of recognition enable us to *know*, argues Butler. However, on occasion, the frame shifts:

> The frame never quite determined precisely what it is we see, think, recognize, and apprehend. Something exceeds the frame that troubles our sense of reality; in other words, something occurs that does not conform to our established understanding of things. (9)

Helplessness in the Bhopal texts is the collapse of frames of recognition and intelligibility (medical, technical-technological, but also political and cultural), leaving only apprehension. Apprehension here functions as the perception of forms of living and dying, but this apprehension does not amount to knowledge: for nobody in Bhopal that night is certain as to *what* they are seeing. In other words, the cultural texts capture the collapse of frames of recognition—of citizens, of forms of life, of medical crises, of possible treatment of the medical crises. The collapse of ways of *knowing* but not *seeing*, documented in the texts, is the helplessness that marks Bhopal that night. Engineers do not know how to deal with the leak, the doctors do not understand the poisoning, the social-political apparatuses of the area cannot comprehend the magnitude or nature of the events. What everybody knows, however, is that the living are *dying*, and they, the ones who ought to have done something to stop this process, are helpless to do so.

"None knew what could be done to alleviate fear and physical discomfort," states Prajapati blandly, about the events that night (26). He adds: "Major Mehra of the Union Carbide . . . knew nothing about the lethal gas and the possible antidote when affected by it" (30). Sharda Dwivedi, who had run around the area to understand what actually transpired, comments to Lapierre and Moro years later about the confusion: "I realized then that something awful was going on . . . something beyond all comprehension" (290). The station master at Bhopal, V. K. Sharma, recalls that, that night, he "felt like the commander of a ship about to be engulfed by the ocean. . . . Even if he could do nothing for the passengers on platform number 1, however, he must still try and save those due to arrive" (Lapierre and Moro 295).

Biological living and dying *matter* is apprehended in Bhopal in accounts of the disaster, but norms that define life—citizenship, primarily—have been effaced, whose status as *life* rather than just living, is obscured.[6]

Butler's focus is on institutional structures of power that determine frames of recognition. My argument here concerns the helplessness generated when the doctors at Hamidia, the station master at Bhopal railway station, the engineers at UCIL, the do-gooders of J. P. Nagar and the colonies are reduced to what they apprehend as living and dying bodies, but cannot recognize their status as *patients* suffering from specific ailments, cannot recognize the rights to medical aid and help which they, as sick, dependent and dying citizens, are entitled to. In other words, helplessness is the result of a failure of all frames of technological-technical, medical and political recognition (which, in Butler's reading involves knowledge, responsibility and even ethics), or rather the slide from recognition to "mere" apprehension. What the eyewitnesses document is a set of affects—primarily horror—and nothing beyond it. There is nothing intelligible about the events they see ("beyond comprehension" as Lapierre and Moro put it), but what they apprehend is brutal, irreducible dying. When recognizability and intelligibility shade into apprehension due to the gas, then bodies are rendered vulnerable to death as never before because there is no social, cultural or medical protection against death.

Helplessness is also the anonymizing of life and death. The individual is stripped, on 2/3 Dec. 1984, of all forms of identity and agency. The gigantic operation of execution—which commentators like Giorgio Agamben have associated with Nazi camps—is the "gas chamber" which is Bhopal (Anees Chishti uses the term "gas chamber" on the first page of his eyewitness account, 25). Helplessness is also, therefore, an anonymizing, the reduction of a human into a mere statistic. Writing about the modern culture of death, Benjamin Noys argues:

> Death is no longer necessarily an individual matter, but now it can be the result of a vast and anonymous operation carried out upon us. In the face of this collective mass death, whatever we have done or might do in life is of no significance, and we are exposed to a death that is "unknown, unhonoured, and unremarked." This exposure to mass death alters our experience of the time of death because death can no longer be the result of our actions, and so we have lost control over death, if, of course, we ever had such control. (2005: 13–14)

Raghu Rai's "Burial of an Unknown Child" (Rai, *Exposure*, www.magnumphotos.com)—the photograph that captured the tragedy of Bhopal and drew the world's attention—emphasizes in the caption (collected in Rai's *Exposure*) that it is an "unknown child." There were three-tier graves because of a shortage of space in the grounds (Rai, photographs titled "Cemetery" and "Burial of Unknown Children" in *Exposure,* unpaginated). Not all bodies were identified, and posters with the photographs of missing people are also part of the visual archive from Bhopal now (Rai, *Exposure*). Piled up bodies are part of the narrative in Lalit Shastri (36) and Prajapati as well (52).

The wide-eyed and panic-stricken doctors at Hamidia also represent helplessness of a different kind: that of the professional unable to make appropriate use of her/his professional expertise. The nurse in *Bhopal: A Prayer for Rain* seeking sodium thiosulphate is horror-struck when she discovers just one bottle in the dispensary. The doctor, watching the crowds pushing, shoving and screaming against the grill gates, stands holding one syringe, with the last of the antidote. After an agonized pan (from the doctor's point of view) of the bleeding, puking and weeping crowds, he reaches for a hand stretched imploringly through the grill, and injects one girl with the antidote. The scene is particularly poignant: one syringe of antidote, hundreds of hands reaching for it, and the doctor, devastated by his inability to treat them all. The doctor too joins the category of the helpless here, precisely because he is equipped to deal with the crisis, and yet not.

Bhopal's helplessness is the destruction of the world as the Bhopalis know it. The world around UCIL will never be a *sustainable* or *life-sustaining* world again because UCIL's processes, people and products have intervened in it and changed it forever. This is so because the effects of the event are not limited to bodies, time or space: they extend across generations, across time and last for ever. Exposed to the gas, the Bhopalis stop being vulnerable, only to be helpless.

The human bodies fall apart, while the material (MIC and the poisons) remain intransigent within the survivor bodies and the environs, a point to which I shall return later.

THE MAKING OF DEATHSCAPES

Scenes of bodies lying across railway tracks, in water canals and small ponds, by the roadside, in their homes and in alleys figure in many of the documentaries and films (shot after the event), and in eyewitness accounts. Raghu Rai's "Mass Cremations" in *Exposure* (Rai, *Exposure*, www.magnumphotos.com) would horrify for its immediacy, just as eyewitness accounts did.

"There were horrid scenes at cremation grounds last evening with flames coming from all directions. In many cases more than one members [sic] of families were cremated on common pyres," writes Chishti, adding in a footnote: "many bodies were buried together in common graves" (38). Such visual and verbal eye-witness accounts of Bhopal from the event (Shastri, Rai, Prajapati, Singh, Chishti) construct a city of deathscapes. Deathscapes intersect with other topographies of sovereignty, memory (in later years, a theme to which I shall return in chapter 5) and work (Young and Light 2013).

Deathscapes are, first and foremost, Bhopal's *geographies of helplessness* and its consequences. Bodies piled outside the hospital and morgue are testimonies to the absence of mechanisms to cope with the disaster. Mass graves are testimonies to the cultural crisis of inadequate spaces and time for death rites. Eyewitness accounts describe how the hospital wards, corridors and grounds outside were filled with dying and dead bodies. The city's streets are, likewise, transformed into giant graveyards and spaces of the dead. These are geographies of helplessness because, as the eyewitness accounts tell us, there was no help to be found in any quarter of the city. The UCIL factory, the railway station, the lake and the hospital were equally incapable (in the first case, intentionally so).

This landscape of dead and dying bodies, the machinery inside UCIL and the environs of Bhopal in the cultural texts brings worlds in pieces into conjunction with bodies in pieces. The faulty machinery inside UCIL produces the faulty bodies outside. Bhopal's abjection is the sum total of expended parts from the plant *and* the human bodies.

The geographies of helplessness transform the city into a space where sovereignty of the human and the institution collapse, where structures designed to offer succor, treatment and support fall apart. In other words, we can read the visual geography in Rai and the verbal geography in eyewitness accounts as mapping the collapse of older geographies of the city (hospital, burial ground, residential area, railway station) and replace them with the new geographies of helplessness. The gas alters, irrevocably for many, the spaces of the city. It is a world now unmade, and which therefore produces the unmaking of subjects/citizens.

Within deathscapes the only identity for citizens is a biological citizenship. Adriana Petryna writing about Chernobyl and its effects defines biological citizenship as "a massive demand for but selective access to a form of social welfare based on medical, scientific, and legal criteria that both acknowledge biological injury and compensate for it" (6). As the effects of the gas become horrifically legible in the immediate days and weeks of the event, accounts such as Rai's and Chishti's turn their attention to the immediate needs of citizen-survivors, and discover the true expanse of deathscapes in the amplification of biological citizenship among them. Injury and death define the people, and the survivors are forced to depend on their own resources.

Sonya Labonte, the Canadian doctor in Rahul Varma's *Bhopal* instructs a victim on the fateful night: "If you're looking for someone, go behind the market square, by the lake . . . there's a big poster on the lawn that says 'Unidentified Dead.' Go there first" (43). Anees Chishti writes:

> Groups of people were seen going round different localities in the old city collecting small charities from townsfolk in bed sheets spread over their heads, stretched from four corners. This indicates the people's disappointment due to the inability of the State Administration to reach necessities to the affected people in this hour of crisis. (32–33)

In the immediate aftermath of the disaster, families went out seeking the bodies of their kin, even as many unclaimed, unidentified bodies were cremated/buried, often in mass graves. In the next few weeks establishing relations with the identified dead became the key effort for many families. Dead bodies, identified or unidentified thus became the nodes in an entire new network of social, institutional and familial *relationships*. These dead were appropriated for affective, cultural and political reasons, as happens with dead bodies all over the world (Verdery 1999, Domanska 2005, Cyr 2013, Young and Light 2013). The deathscapes of Bhopal, in other words, generate a set of social and familial relationships founded entirely on injury and death.

The visual archive of mass graves and heaped dead bodies offer, in the immediate aftermath of the disaster as documented in the accounts reporting the events, two major forms of evidence (Domanska 2005). First, they constitute evidence of the technological, cultural and juridical collapse that is Bhopal (it is juridical for the simple reason that it has been addressed frequently as a *crime*). Second, they function—and continue to function even today—as the reference point for the work of mourning. If the first is couched in the rhetoric of justice, the second is cast in the rhetoric of memory (Domanska 401).

Lalit Shastri, T. S. Chouhan and Prajapati point their fingers at the UCIL's history of negligence and indifference, and the Indian medical services' complete failure as the cause behind the massive numbers. Shastri, for instance, notes how, first the UCIL authorities did not offer any clear information about the constituents of the gas that leaked, nor its effects. The Indian Council of Medical Research's (ICMR) studies regarding the gas and its effects were not made public—I shall return to this theme of secrecy in a later chapter—and "the flow of medical information virtually stopped" (55). Then Shastri writes: "*in these circumstances* deaths continued unabated and each day the number of the dead mounted. Week after week, new patients, flocked to the hospitals" (56, emphasis added). The mounting number of bodies in various alleys and public places become evidence, in Shastri's account, of the technological and social apparatuses of Bhopal city. The mass graves are evidence of both a crime (by UCIL) and of apathy (of the Indian organizations, such as ICMR). Abject bodies are not the end-product of biomedical crises alone: they are the result of the thanatopolitics driven by both biomedical "reason" and neoliberal policies that enabled UCIL to get away with their actions, from the moment of their establishment of the plant.

Images of mourning families around mass cremations and burials transform the mass graves into the reference site for mourning, for collective memories. Rai's photograph of sacks full of skulls constitute, likewise, a symbol of the reminders and remainders of the events.

The visual and verbal accounts of mass graves and cremation also serve a witness-function (Cyr 2013). The stacks of skulls in Rai's photograph (Rai, *Exposure*, www.magnumphotos.com) provide the evidence for the uncounted and often unidentifiable individuals who died in the disaster. As a component of the new geographies of Bhopal, these sites and visual records such as the above-mentioned photograph from Raghu Rai, are evidentiary sites. They constitute a witness-function even when they are no longer witnesses in the full sense of the term. That is, they are *material* witnesses for the events that have occurred.

Finally, there is an unfortunate deindividualization of the person in such events as Bhopal that all eyewitness and retrospective accounts describe. Diana Kontsevaia (2013) has argued that mass graves deindividualize the person/victim, but also enables her/his insertion in death into a grander, collective narrative. "Many of those who had collapsed and died while fleeing from the affected areas during the early hours of the morning were lifted and their bodies disposed of at the state initiative and no trace was left of them," writes the eyewitness Shastri. Those missing were never traced, those buried were never identified. Shastri notes: "it was a common sight to see grim and saddened faces peering at posters showing the photographs of the unclaimed bodies searching desperately for their friends and loved ones" (56).

Deathscapes also, however, must be taken to include the landscape of the missing and the disappeared because these have died *social* deaths—never identified, never recovered, and never acknowledged by the state apparatus.

Helplessness is also the helpless ignorance of the survivors, who do not know what happened to their loved ones and acquaintances. This too, again, gestures at the failure of cultural protection when no machinery was available to document and identify the dead, reducing Bhopal to an *anonymizing* geography of the dead. In the Bhopal Gothic, part of the secrecy remains, in terms of this anonymity, of the missing and the dead. Precarity consists in continuing secrecy and anonymity as well.

NOTES

1. It should be emphasized that I am not employing "thanatopolitics" in the current sense used by Roberto Esposito and others, wherein the term signifies populations identified as unhealthy or unhygienic are marked for containment or even extermination so that the rest of the population stays healthy. I use thanatopolitics to signify a state of affairs where ill-health and suffering in an entire section of the population—the gas exposed—is ignored so that their deaths simply do not count.

2. The Gothic genre "abounds in images of . . . the corpse, whole and mutilated," and of "an array of bodily wastes such as blood, vomit, saliva, sweat, tears and putrefying flesh" (Creed 1993: 10). Further, "an 'abject' also resembles but is not quite an 'object.' The content of the 'abject' is always loathsome," writes Kelly Hurley (2007: 144).

3. The "vulnerable being," Cavarero writes, "is here the absolutely exposed and helpless one who is awaiting care and has no means to defend itself against wounding" (2011: 20). The helpless is in a "condition of passivity, undergoing violence he can neither flee from nor defend against" (30).

4. See Susan Derwin on the animalization of the human in Nazi camps (2012).

5. Murphy writes: "a human body is by definition vulnerable, and this ontological fact cannot be parsed from the ethical obligations that it suggests. . . . The vulnerable human body is also the provocation for an ethics insofar as it elicits a response" (2011: 577).

6. To cite Butler, "when those frames that govern the relative and differential recognizability of lives come apart—as part of the very mechanism of their circulation—it becomes possible to apprehend something about what or who is living but has not been generally 'recognized' as a life" (13).

Chapter Three

Bhopal's Biopolitical Uncanny I

The Nature of Haunting

There is a Bhopal uncanny. It is marked by secrecy, repetition, the spectral, by the interface of the familiar with the strange, by simultaneous secrecy and revelation. The Bhopal uncanny is effected through the transformation of the familiar space of the UCIL site and the environs into an unrecognizable and threatening landscape on 2/3 December 1984, and forever afterward. But it is not the gas leak alone that characterizes the Bhopal uncanny.

This chapter, the first of the two on haunting in Bhopal's ecological Gothic, has three subsections. In the section on the repressed, I examine the secrecy and misinformation around the plant, its processes, safety measures and products that constitute the ecosystem of the world of Bhopal. (Incidentally, the theme of secrecy and misinformation has surfaced time and again around UCIL's West Virginia plant, as the 1991 documentary *Chemical Valley* reports.) It is this secrecy, the repressed, that remains the central figure in the haunting of Bhopal's citizens. Even 30 years after the disaster, as some campaign materials point out, the government of India claims that there is nothing to hide in terms of the plant's toxic wastes, since there are no wastes (*25 Years after Midnight*, 1991). N. R. Bhandari writes: "Union Carbide continues to treat medical information on the leaked chemicals as a trade secret. Till date very little is known about how to treat chronic exposure gas induced diseases" (2010: 35). The secrecy around the plant remains unaltered, even as poisons are secreted into the water resources around the area. As Anjali in *A Breath of Fresh Air* puts it, "I would always vent that no one had bothered to tell me that the Bhopal gas tragedy had left its mark on my womb. I wish someone had told me that having a child would be dangerous to the child, that any child I had would be affected by that fateful night. . . ." (93).

In the section on the monstrous I pay attention to the effigies and posters that clearly treat the local disaster of Bhopal as embodying a global process, and the unnamed, unidentified but palpable "presence" of global power. The monstrous in the Bhopal representations is not the patriarch or avuncular guardian of the traditional Gothic but rather the state (or Government of India) who stands in the place of the guardian. The Bhopal protests repeatedly reinforce the claim that the state as protector and as guardian has not only failed in its responsibility but actively supported the wrong-doers. The ecological Gothic's monstrous does not have murderous uncles or incestuous fathers: what it has, instead, is the state in the role of *paterfamilias* to whom much of the evil might be traced. In the section on "In-corporations and Unholy Liaisons," inspired by the effigy where Dow Chemicals is represented as a demon with 10 hands, of which several are Indian and some American, I study the disaster's cultural representation of alliances and hybridization that contributed to 2/3 Dec. 1984. Finally, the chapter turns to the representations of the disaster as an autoimmune event.

APPREHENSION, RECOGNITION AND THE REPRESSED

The Bhopal events appear as the uncanny in the cultural texts. Information and knowledge about the UCIL operations and the products it manufactured were secreted away in files and communications. Further, the antidotes and safety measures were unavailable to the people who needed them the most. Finally, the available information and documentation was both incomplete and misleading. UCIL rests on repressed materials: specifically, information. Admittedly, this interpretation emphasizes a textual uncanny—secret information and ambiguous information—undergirding the Bhopal disaster. However, such an interpretation is invited by the representations available to us in the form of both eyewitness accounts and later-day commentaries of the events.

The Bhopal uncanny is constituted by the "condition" of information and knowledge, both of which are repressed in specific ways. I propose that this condition of information is in fact a gap between apprehension and recognition.

Apprehension, writes Judith Butler, "is less precise . . . it can imply marking, registering, acknowledging without full cognition . . . a form of knowing . . . bound up with sensing and perceiving" (2009: 5). "Recognition" implies norms of recognition, "more general conditions, historically articulated and enforced, of 'recognizability'" (5). Recognizability implies conditions that "prepare or shape a subject for recognition" (5). It is linked to "intelligibility," "the general historical schema or schemas that establish domains of the knowable" (6). In the Bhopal cultural texts we see this gap between sensing and knowing, between apprehension and recognition, represented as insurmountable.

In the movie *Bhopal: A Prayer for Rain*, a worker tells a doctor that ever since he started working in UCIL his health has been steadily deteriorating. In the same film, Dilip, the newly hired worker, begins to suspect that things are unsafe and in his sneak call to the journalist Motwani, admits that he is afraid. The incidents involving injuries and deaths of co-workers clearly prey on their minds. This is apprehension: the *sense* that something is wrong. There is, however, no way of ascertaining whether their apprehensions are founded on anything concrete. There is, in other words, no historical, empirical or even documentary schema by which Dilip and the workers can *understand* the UCIL operations. This gap between apprehension and recognition is brought home to us in a particularly powerful scene in this film.

The doctor to whom the worker has complained of ill-health, responds in a significant way. With a supercilious air, he waves a printed *English* pamphlet at the worker and tells him, "It is not that all the chemicals in UCIL are harmful to humans." The worker clearly is unable to read the pamphlet and relies on the doctor's "superior" wisdom regarding the matter. The pamphlet that the doctor opens out has a section titled "Recommended Hood Use," printed in large-sized font. The smooth misdirection of information is startling: the worker's query about the link between the conditions of his workplace and his deteriorating health are answered by the doctor at two levels:

1. The chemicals are not all toxic (addressed to the worker)
2. Retaining the hood as a safety mechanism is the worker's responsibility (addressed to the viewers)

Later in the film the safety engineer Roy explains the procedures to neutralize MIC, in the event of its leaking, to Dilip. He particularly warns Dilip that MIC and water are never to mix. Dilip calls the journalist Motwani (the film version of Keswani, who carried out the numerous exposés on UCIL, only to be ignored) and reports what he had been instructed: "something about MIC and water, about mixing. . . . I don't know, I didn't understand." This absence of understanding of the proper procedures is at the heart of the uncanny: Roy is unable to explain the procedures to the illiterate worker.

Information regarding the chemicals is not supplied to the worker, nor does the doctor admit to any responsibility on the part of UCIL. Instead he directs attention, via the pamphlet, to the *worker's* own responsibilities. Thus, information of one kind that would be crucial to admitting UCIL's responsibility is denied, while another which foregrounds the responsibility of the worker is supplied. This information, which would have served the purpose of recognition (and appropriate response), is cast in a language the worker does not understand. The worker, like Dilip, *senses* that his ill-health is directly connected to the UCIL plant and its chemicals, but he is denied intelligibility because frames of interpretation are kept secret from him. Further, with the doctor alone in charge of these frames, the authority of interpretation is solely his. The sense of something wrong comes up against this authority and its frames of recognition, and collapses.

The absence of appropriate frames of recognition generates the secrecy around UCIL. Secrecy and uncertainty are, in Freud, integral to the uncanny (1971: 370, 373–75). Maria Tatar suggests, following the work of Tzvetan Todorov, that the ambiguity in the uncanny event "generates the hesitation that defines the fantastic" (1981: 169). There is both hesitation and "intellectual uncertainty," writes Tatar (169). What is significant in Tatar's reading of the uncanny is the emphasis on "hesitation" as the state of mind of the perceiving subject. The uncanny emerges over a period of time rather than in an instant when the "secretly implied meanings in a text" intrude, as Christine Berthin defines the uncanny (106), into the triumphalist text of Bhopal's narratives resulting in ambiguity.

It is possible to also see the secrecy of information and the uncertainty resulting from this as the failure of cultural protection: the workers who are the key operators within the process do not recognize what they handle at such risk. This too is a sign of the Gothicization of Bhopal. Gothicization, as argued earlier, links humans, organic nature such as water, plants and land, and the nonhuman chemicals such as MIC or phosgene). If the process of Gothicization is the iteration of material bodies and their insistence within networks that involve corporate bodies, the human and the nonhuman, then the information exchanges necessary between bodies (engineers, workers), corporate bodies (UCIL's administrative sections) and the chemicals themselves do not occur. Cultural protection which would have resulted from this exchange, therefore, does not exist at all. The "crumbling" and ruination associated with the Gothic (Wasson) is the erosion of trust, information exchange and chain of command inside UCIL over a period of time.

The "uncanny" is linked to the word "canny," which implies knowledge and recognizability based on frames of reference and intelligibility. But etymologically "canny" is also linked to "occult" or "supernatural" powers, as is its very opposite: the "uncanny." Tatar tells us that "what is canny can easily become the uncanny" where the "un" can "figure as a token of repression" (170). There is one more dimension to Tatar's reading of the uncanny that serves our purpose here. Tatar notes that Freud's *unheimlich* (the uncanny) is linked to the home/house. A house "contains the familiar and the congenial, but at the same time screens the familiar and the congenial from view, making a mystery of it" (169). Tatar continues: "what takes place within the four walls of a house remains a mystery to those shut out from it" (169). The opening paragraph of *The Bhopal Reader* states:

> As people woke up, gasping, choking, coughing, nobody knew which way to run to escape the lethal effects. Nobody knew that it was better if you didn't run from the burning fumes, that the harder you breathed, the worse it was for your body. Nobody knew what was leaking from the Union Carbide pesticide plant, most did not even know what was produced there. (Hanna et al. 3)

Dr. Kumkum Saxena, who was a medical officer with UCIL, says in *One Night in Bhopal*, "I did feel this was the Carbide family." And yet in the same film we hear a comment: "Despite their connections with the factory, Shahid and his family knew nothing of the hazardous chemicals." Very few of the workers in UCIL had any real idea of what the company manufactured. Indeed, many believed that UCIL "manufactured medicines for plants," says the commentator in the film. That is, even to people involved in the day-to-day operations of the place, to which they believed they belonged as members of the UCIL "family," there was no access to knowledge that would have saved their lives and that of others. They *feel* something is "off" but have no way of confirming (or rejecting) this sense.

In Rahul Varma's *Bhopal*, Devraj admits to the UCIL chief, Warren Anderson, that Sonya Labonte's research into the effects of UCIL's chemicals had been secreted away: "she was doing research claiming that Carbide International had been poisoning unborn babies ever since its Bhopal plant was built" (2004: 50). In the first minute of *One Night in Bhopal* we are given by voiceover some statistics: "8,000 killed and over 200,000 injured." This is interrupted by another voice which says, "the company says it doesn't know what caused the accident." N. R. Bhandari, medical superintendent of Hamidia Hospital in 1984, records in his eyewitness account: "I was informed the gas is PHOSGENE which is a toxic gas. No guidelines, no information about antidote, no information about the effect of the gas on the human body were intimated" (2010: 13).

Lapierre and Moro record that UCIL never shared any information about MIC with the city's Bhopal's medical corps (304). Other reports from investigations into earlier accidents were not made public (D'Silva, "The History of Accidents at the Bhopal Plant"). In the preface to Chouhan's work, Kim Laughlin states:

> A hard, green crust had appeared on the surface of stored food but people were told that there was no "scientific reason" for concern that it was contaminated. A helpless dread spread throughout the city, with people very conscious of their lack of control over the continued risks. . . . When victims approached the Carbide factory for medical aid and information, desperate managers afraid of rioting suggested that another leak might soon occur. (8)

Laughlin continues:

> Morcha activists were completely barred from official information sources. The clamp-down reached from the highest city authorities to the ward nurses at government hospitals. . . . As one activist explained, "from long before the disaster there was a manufactured silence that barred any preventative action." (9)

During the years of operation, the "text" of UCIL history was a device for concealing secrets. Chouhan writes:

> Many other leaks also took place in the MIC plant in 1983 and 1984 . . . workers would hurriedly leave and all work would stop for some time. Injuries during these leaks were treated at the plant dispensary, with incidents usually *recorded as cases of material loss rather than injuries to workers.* (1994: 37)

Lalit Suri in his eyewitness account notes how the doctors at the Hamidia hospital queried the medical officer of UCIL and other employees for any information regarding the nature of the gas leaked, and were provided either contradictory information or no information at all (42–43). "It soon became evident," he writes, "that no one knew what line of treatment to adopt" (43).

Recognition implies acceptance and knowledge of schemes of interpretation. When UCIL denies crucial information about the dangers of MIC, or even the history of accidents—blaming worker negligence for these—the frames of recognition are tampered with. Apprehension does not stand a chance against the stout denials by the UCIL authorities. Recognition being organized by and in authority, the words of the chairman or the higher authorities in these cultural texts serve as assertions of *knowledgable* fact.

Bhopal's uncanny stems from the condition wherein its true nature—its chemicals, the toxicity and antidote information—lies concealed all through the plant's existence. The plant, in other words, does not offer itself up to clear vision. It lies partially secreted. There is a sense in the uncanny of the "secret encounter," an "apprehension . . . of something that should have remained secret and hidden has come to light" (Royle 2003: 2). In the Bhopal texts the plant's secrets are encountered in the form of the accidents and the horrific deaths of Ashraf and other workers *before* 2/3 December, it is *sensed* (apprehended) but not always seen or *understood*, like a secret glimpsed at but not fully revealed. Journalists like Keswani who repeatedly warn of the impending disaster—because he reads the signs carefully—are ignored. It is at once, then, the denial of knowledge by UCIL and the rejection of knowledge by the neighboring people and the workers themselves that would remain to haunt Bhopalis for ever after. The Bhopal uncanny is founded in not false or muddied perceptions but in epistemological uncertainty about the plant.

The "hesitation" (Tatar) or "undecidability" (Weber 1973) that marks the uncanny may be found in several cultural texts on Bhopal. (There is another dimension, or aspect, to this undecidability theme, in connection with the Bhopal uncanny, to which I shall turn later in this chapter.) First, the control room operators, as shown in all the films, cannot believe that the meters and monitors in the plant are reflecting the true state of affairs of E610. Second, once they acknowledge that things are going wrong, they have no solutions to the problem because they have not been adequately briefed, trained or prepared. Third, there is ambiguity and ambivalence about the nature of the gas, its antidotes and the treatments. On the face of every UCIL worker in these films we see the tension between the illusion of knowledge and the reality of ignorance. The illusion of knowledge proceeds from the assumption that UCIL was the safest plant in the world. Lapierre and Moro record Qureishi's disbelief at the collapse of this illusion because her was "convinced that no accident could happen in a factory that was not running" (276) and had implicit "faith" in the MIC tank to resist stresses and strains (277). Ignorant as they were of the exact nature of the safety measures that had been *ignored* by the company earlier in the week and that night, the supervisors on duty on 2/3 December 1984 had no idea as to how to restore the balance.

The hospital scenes in the films, likewise, point to the hesitation-undecidability-uncertainty dimension of Bhopal's nightmare. The doctors, working with no knowledge of the constituents of the gas or its antidotes, frantically call UCIL for information, and it is not provided. The consequences are horrendous: no doctor has anything other than symptomatic treatment for the thousands streaming into Hamidia Hospital.

> As people lay dying Union Carbide's Director of Health & Safety, Jackson Browning, said that methyl isocyanate (described in UCC's own safety literature as a "poison by inhalation . . . major residual injury is likely in spite of prompt treatment") was "nothing more than a potent tear gas." (22)

And here is the story "Lie and Let Die" in *The Bhopal Marathon*:

> Union Carbide's internal safety manual, written in 1974, stated that if inhaled MIC could cause "fatal pulmonary oedema" and that "major residual injury is likely in spite of prompt treatment." So dangerous was MIC that UCC gave it the maximum rating possible in its internal hazard system. (22)

Information about the actual constituents of the gas were not shared because, as Union Carbide declared, these were "trade secrets, which are therefore not available" ("Lie and Let Die," 24). It is in this gap between reality and misinformation, knowledge and ignorance that the Bhopal uncanny asserts itself. (This story in *The Bhopal Marathon* lists the letters, memos and other documents through which the silencing of information was effected and available information contradicted by UCIL.) Records of the first meetings between Dr. Max Daunderer, a German toxicologist sent by the Government of India to help in the treatment of victims, and Dr. B. B. L. Mathur, the dean of Gandhi Medical College, Bhopal, states clearly that sodium thiosulphate was a prophylaxis that was administered and patients "showed overall improvement as per Prof. Bhandari" (APPEN, appendix 5.6, 111). The minutes then declare that "sodium thiosulphate was not at all harmful." Later UCIL and a few doctors would deny the efficacy of sodium thiosulphate. Within days, a "Confidential Letter" of the same Prof. Mathur refers to a letter received from Dr. K. G. Ragu, director, Medical Services, Bhopal, on 13 December 1984, states, "under no circumstances Antidote sodium thiosulphate be used in any case unless absolute positive and correct sign of cyanide is found in blood samples" (APPEN, Appendix 5.9, 121). This reversal of what would turn out to be a key factor in the continuing suffering and deaths of thousands who inhaled MIC that night, had no scientific basis, as "Lie and Let Die" states (25). N. R. Bhandari, the doctor at Hamidia, writes in his memoir:

> Ten days after the disaster, Union Carbide Corporation's medical director first supported mass administration of thiosulphate and, in another telex message three days later, forbade it. . . . Director of Health Services, Dr MN Nagu sent a circular to all doctors in the city warning them that they would be held responsible for any untoward consequences of thiosulphate administration. . . . Interestingly, given that no adverse impact of administration of thiosulphate was reported in literature, the circular had no medical basis. (2010: 57)[1]

This imperative was accompanied by completely misguided instructions to the people. Thus a senior police officer announced that "the effect of the gas had subsided and people should return home." This, writes Moti Singh, Collector and District Magistrate of Bhopal in 1984, was "a very uncalled for and erroneous direction" (2008: 35). This is another instance of misinformation.[2]

The thanatopolitics around gas victims—the lack of information from UCIL or UCIL, the absence of proper information within ICMR and the results of the various studies—demonstrates a nexus between biomedicalization and the neoliberalism that enabled UCIL to set up shop in India/Bhopal in the first instance. That is, the biomedicalization of the disaster and the "victims" is linked inextricably to the freedoms granted to UCIL to conduct its production processes with no responsibility towards the local population and the risks they (the Bhopalis) lived under, as all the texts indicate. The contradictory messages regarding possible antidotes is not merely an instance of skewed biomedical advice: it has everything to do with UCIL's clout in ensuring that the true nature of the poisonous gas—with elements of cyanide—is never revealed.

It is knowledge that both constructs and destroys the boundary between canny engineers or doctors and the uncanny experience of 2/3 December, between the secret knowledge within the home (UCIL, USA) and the ignorance that was writ large on that night. Doctors, supervisors, engineers and administrators find themselves locked into a bureaucratic fight with the UCIL authorities who are reluctant to release crucial information. To extend the binary of canny knowledge and uncanny experience further, it could be argued that the UCIL home/house kept its secrets, even from its inmates. Thus, what was familiar—the factory premises, the control room, the processes and even the chemicals themselves—is rendered strange, although the strange was always part of the familiar, given the secrecy around the MIC. In other words, even to the most professional engineers and workers the plant seemed to *become* strange that night, if the expressions on the faces of these characters are anything to go by.

I term this the Bhopal uncanny because that night UCIL reveals its secrets, its essence, so to speak, and thereby defamiliarizes itself even to its most intimate worker and neighbor. Take, for instance, the shock on Suman Dey's face when he discovers, in *One Night in Bhopal*, that the hoses do *not* reach the top of the chimney/vent. This knowledge, combined with an awareness that the gas spewing out of the chimney is one of the most potent chemicals ever made, renders the plant something strange: it is no longer for Dey or Qureishi the plant they (thought they) knew. Their ignorance of processes and flawed mechanisms renders the place uncanny: at once familiar and strange.

If we consider, following Freud, the uncanny to be primarily the effect of a place on perceptions, then viewing the reconstructions of disaster also induce in us the uncanny.

Documentaries made after the disaster linger on the broken down, abandoned plant and its parts, rendering these strange and unfamiliar. Yet, this strangeness also occurs in the fictionalized reconstructions of the events of 2/3 December 1984. There is no mimetic exactitude in the reconstruction of the events, and we have to assume that "this is how it was," if necessary by comparing it with eyewitness accounts. As Rahul Mukherjee (2012) has argued, present-day "situated testimonies and the presentation of archival footage actually supplement and nourish each other" (7), even though this implies a selection of footage (only that which suits the story being narrated *now*).

In the films, we see Suman Dey and his team try to use the water hose on the gas tower emanating the poisonous gas, to no avail. We see the engineers battle the tank, the gauges and the vent gas scrubber (VGS). I see this re-enactment of events as contributing to the strangeness of the place because it transforms the events from long ago into something familiar—as in visible, legible—and yet, because these documentaries also offer us views of the present-day ruined plant, as strange. Since we watch these scenes of re-enactment of historical events through carefully spliced archival footage (such as photographs of the dead and mass cremations in Prajapati, or Bhandari, for example, TV grabs and yellowed photographs of hospital scenes, or people carrying off bodies on 3 December in *Bhopali*) we are drawn to the UCIL site as a landscape at once familiar and strange. The landscape is an uncertain one, with the atmosphere (and atmosphere is central to the uncanny as Freud showed) moving between efficient workings and nonworking, changing in form and function from productive space to killing field. It is possible to see, with the historical awareness (of previous accidents, of the history of negligence discussed in chapter 1) imparted to us by the cultural texts themselves, that our viewing of the reconstructions is itself a form of uncanny repetition of the events: we watch the events of 2/3 December 1984 which themselves repeat, on a larger scale, similar accidents from earlier years of UCIL.

It is the repressed that results in the Bhopal uncanny: the information secreted away and which results in ambiguity, uncertainty and eventually catastrophe. MIC is the Thing, the secret at the heart of UCIL's uncanny. It embodies the repressed information that UCIL never shared. The Bhopal uncanny is the effect of the ambiguity, contradiction and secrecy around information and texts generated by UCIL in its history. The site of the factory, the gas itself and the symptoms are open to multiple and ambiguous interpretation, as the commentaries and eyewitness accounts show, because there is a history of textual obfuscation in UCIL and the state as well. Thus, in 2009 Shivraj Singh Chohan, Chief Minister of Madhya Pradesh in a BBC interview, incorporated into the documentary *Bhopali*, flatly denies that there is any remaining waste from UCIL anywhere in Bhopal or its environs. The next frame of the documentary shows Chandra Bhushan, an associate director of the Centre for Science and Environment, offering detailed analysis and reports of the pollutants and contaminants in the water and soil of Bhopal.

There is one more aspect to the Bhopal uncanny and its repressed. In alignment with the traditional definitions of the uncanny, the spectral and the past haunts the present in the cultural texts. Every film and text refers to the dead Ashraf and the history of accidents involving MIC and phosgene in the UCIL plant. With every accident in the sequence there is a return to a similar accident in the past, almost as though the past unsettles the present as well. It is the repetition of the accidents that constitute the haunting of UCIL and Bhopal. History in the case of UCIL and Bhopal is not linear, but cyclical. Events from the past repeat, endlessly. The repressed returns, even when Satinath Sarangi, founder and managing trustee of the Sambhavna Clinic shows us the clinic's premises in *Bhopali*. At one point, he gestures at the "top of the MIC plant structure" from where the gas leaked. He tells us: "it is only about 400 or 500 meters from here." Sarangi's comment suggests that the MIC unit, like UCIL, is integrated into the premises, lives and futures of the people even today.

This uncanny knowledge and information transforms Bhopal into a Gothic state on 2/3 December 1984. Writing about the link between information and the Gothic Victor Sage says: the "mechanical withholding of information . . . becomes in the context of the Gothic a metaphor for epistemological doubt and theological unease in the presence of death" (cited in Horner and Zlosnik 2012: 327). The withholding of information was, as Kumkum Saxena discovers when she works for UCIL (*One Night in Bhopal*), a matter of routine "management." Her discoveries of the toxicity of MIC and its risks are dismissed by the higher echelons.

A further point may be made about the tense gap between apprehension and recognition. The documentaries and feature films about Bhopal offer us glimpses of, and often direct commentaries on, the sense of something wrong vis-à-vis the technical and safety aspects of the UCIL operations and the frames recognition that are grounded in financial matters and profits. Roy in *Bhopal: A Prayer for Rain* senses the potential problems in abandoning safety measures in favor of cost-cutting, and as we can discern clearly from his expression when his objections at the shutting down of the MIC air-conditioning unit by Chaudhury the supervisor, are overruled. Chaudhury, overruling Roy's objections, asks him if he has any idea of the costs of air-conditioning. In *One Night in Bhopal* Kumkum Saxena conveys her sense of something amiss when she takes her toxicology reports to her boss, who scoffs at her perceptions of risk.

In both these instances, the protagonists *visibly* convey unease and distress: they *sense* that something is wrong when safety precautions are ignored. This is their apprehension that UCIL is not working as it should. However, the hard evidence of financial losses that the accountants counter their objections with is a clearly identifiable frame of recognition. The shift from unnamed and unnameable fears to numbers, currency and sales graphs is the shift from apprehension to recognition. UCIL, as we see it, has the frames of recognition that deny apprehension and the sense of something wrong, any validity.

Repression in the context of the Bhopal uncanny, then, is the suppression of fears and anxieties and their subsumption under frames of recognition such as sales and profits. But repression also operates at another level wherein crucial information is kept from workers, administrators and the general public. We therefore need to see Bhopal's uncanny as the product of a double repression.

If the uncanny implies the revelation of what should have remained hidden, then 2/3 December 1984 reveals the secreted information regarding MIC. It brings to the surface the falsehoods and misinformation but also the downplaying of apprehensions of risk and the highlighting of the profitmotif. The precarious culture of Bhopal and its aftermath is the collapse of available frames of knowing.

THE MONSTROUS UNCANNY

Bhopal's uncanny haunting involves a monstrosity, which is at once home-grown and global, both internal and external, so that it becomes difficult to isolate either as the primary source of the events of 2/3 December 1984, as the cultural text demonstrate. The monster/monstrous, as commentators have argued (Cohen), was one that existed on the peripheries of the known world, and invaded the human habitations to produce destruction. Beyond the boundaries of the known world existed lands and life forms that did not obey known laws. The monster always came in from the outside, breaching borders and perimeters.

Nearly a decade before liberalization becomes official economic policy in India (it is dated, most often, to 1991), the Bhopal disaster represents the erosion of these perimeters beyond which monsters existed. Bhopal's tragedy is the effect of the monstrous link between internal and external forces, a feature of the world we have only *now* come to associate with globalization.

In his essay on globalization and disaster, Walter Kalaidjian writes: "not simply an effect or cause of globalization, disaster happens more as the former's internal limit: one that demystifies globalization as itself being the disastrous object" (2007: 829). Kalaidjian goes on to argue, via W. J. T. Mitchell, who in turn is following Jacques Derrida, that a disaster like 9/11 is indicative of globalization that is somehow homegrown. It is this line of thought I pursue in order to read Bhopal's tragedy as an effect of the monster within *and* without: a monster generated as a result of a miscegenation between the local and the global. The "monster" is not a creature in this reading: rather, it is an unnameable Thing, an effect, at the heart of the disaster.[3]

Satinath Sarangi introduces UCIL in *Bhopali* thus: "Union Carbide Corporation is this American transnational that was involved in the world wars." But UCIL's monstrosity was not the result of its American, transnational or capitalist *origins* alone: its monstrosity, as the cultural texts demonstrate, is a hybrid.

The Neo-imperial Gothic, In-corporations and Unholy Liaisons

UCIL as a foreign body and presence that has invaded, and been incorporated into, the social fabric, the environment, the event and the psyche of Bhopal/Bhopalis might be read within the framework of the Gothicization of urban space as well. Kathleen Spencer in her work on Bram Stoker's "urban Gothic" in *Dracula* noted how late-Victorian authors set their Gothic fantasies in urban settings, and used the language of scientific empiricism to describe the spaces (Spencer 1992). Patrick Brantlinger has noted that such accounts of Gothic London encoded a fear of foreign invasion and consequent racial collapse of the British—what he termed the "imperial Gothic" (Brantlinger 1988). In the case of Bhopal, the invasion and intrusion is imperial too—but a neo-imperial one, of American companies of/into the postcolonial nation-state. Thus, the neo-imperial Gothic is a 20th-century *reversal* of the imperial one, where the colony invaded the (white, European) metropolis and nation.

The foreign body that disrupts the postcolonial nation-state and induces collapse and almost perpetual decay in its body-politic is UCIL. It infects, poisons and induces permanent mutations. The UCIL-as-foreign body is however not acting only as an invasive agent: it works through an intricate, near-symbiotic relationship with the local structures and body politic, made possible by in-corporations and liaisons. I use the hyphenated form of the term "incorporations" to draw attention to the invasive, intrusive and assimilatory dimension of the "in," suggesting the internalization of the foreign, but also to signal that the foreign is a *corporation*.

In Varma's *Bhopal* it is the unholy liaison between Carbide's profiteering and India's corruption that collectively produces the tragedy. The Anderson-Devraj Sarthi exchange in scene 15 (47–51) clearly indicates this. Anderson denies knowledge of the way the Indian plant operated. Sarthi informs him that he, Sarthi, acted on instructions from the American headquarters to cut costs and maximize profits. N. R. Bhandari puts it even more bluntly in his memoir: "it seems that they [Indian government officials and safety inspectors] were drawing salary from the Government but working for Union Carbide" (102). In an early critical essay media theorist Arvind Rajagopal stated:

> In a situation where Third World countries compete with each other for the attention of multinationals, strict regulatory action could be seen as part of an unfavourable investment climate. In all of this, the deep ideological commitment to "modernization," and to emulating the example of the West, cannot be underestimated. (1987)

Rajagopal not only draws attention to the necessity of Third World nations to allow the Carbide kind of exploitative investment he also demonstrates how a model of modernization adapted from the West is readily implemented in Third World conditions. This is an instance of the monstrous hybrid as well.

In chapter 3, I argued that the effigy when burnt reveals its innards, and the boundaries between inside and outside are burst wide open, indicative of the abjection that is Bhopal. However, there is another way of perceiving the effigy burning.

The three effigies instantiate the Bhopal uncanny's monstrous, in particular its hybridity. In the effigy from 2003, we have a reworking of the very recognizably Indian Dussehra festival when, on the 10th day, the effigy of

Effigy from 2003: Protests. Courtesy International Campaign for Justice for Bhopal.

10-headed Ravana, the demon-king of Sri Lanka, is burnt. Ravana is an embodiment of evil power according to Hindu texts, and the burning recalls the victory of Rama, the god-king, over Ravana in the *Ramayana*. Ravana's

Effigy from 2004. Courtesy International Campaign for Justice for Bhopal.

Effigy from 2007. Courtesy International Campaign for Justice for Bhopal.

10 heads in the Bhopal protests are hybrids: they represent Dow, the chief minister of Madhya Pradesh, the Indian government and UCIL. The 10 heads, it should be noted, belong to one body. In the second effigy, the Indian government and the US government are united in the body of UCIL, which

wears the hat from Dow. In the third one, from 2007, the effigy has multiple arms and is two-faced. The arms are labeled after various industrialists and politicians: P. Chidambaram (politician), Ratan Tata (industrialist), Vilasrao Deshmukh (politician), Shivraj Singh Chauhan (politician), Andrew Liveris (CEO, Dow Chemical) and others.[4]

These representations thrive on the amplified grotesquerie of hybrid bodies—the staple of Gothic and horror fiction (Hurley, Morgan 2002), and commonplace in Hindu mythology. We can discern several layers to the symbolism here.

Most prominently, the effigies convey a sense of unity, of process and purpose, among the Indian politicians, businessmen and American corporate leaders. They all occupy one *body*. The effigies convey the sense that the processes commonly associated with the biological body are all organized within the same space. They are all integrated into one process, and the impression of the coherently working body is what stays with us.

The effigy also symbolizes the coexistence within one corporeal entity, of many life forms, species and identities: Dow, UCIL, the Indian state, industrial tycoons. Bhandari speaks of the "[Indian] state's implicit support for the offending corporation while practicing disdain for the victims" (33) . The corporation, we see, has been in-corporated into the *body* which is then treated as the monster by the gas victims. This is the body of monstrous action in Bhopal. The biomorphic symbolism of the effigy calls upon us to imagine the victim and the victimizer as *bodies*, whether flesh-and-blood bodies or corporate bodies. The monstrous is the mix of these two within the biomorphic representation of the effigy.

The effigy incorporates the other into the self, the foreign into the self. The self fragmenting into the Other is a traditional Gothic trope,[5] but the Bhopal effigies suggest that the *reverse* of this fragmentation can be equally nightmarish: the "Indian" company, the plant, the site and the process now *includes* a foreign body, has a foreign body merged/fused into it. This of course brings it in line with the uncanny, which, as Nicholas Royle argues in his magisterial work, "may . . . be construed as a foreign body within oneself, even the experience of oneself *as* a foreign body" (2). UCIL is at once recognizably Indian *and* foreign. We see the foreign in the bits-and-pieces form of the effigy: which is one and yet not just one.

This hybrid corporeality, if that is what it is, of the effigy generates the uncertainty and undecidability that marks the uncanny. There are uncanny resemblances, in the multiple faces/heads on the effigy, to recognizable figures and real persons, like Warren Anderson or Chidambaram. Yet, their location in a body that is dressed in an all-black suit or with an American flag or with a top-hat imprinted with the Dow logo renders them unfamiliar. This uncanny resemblance is linked to the Bhopal uncanny, which is committed to mourning the dead of the tragedy. The interesting conundrum, or the uncanny, here in the case of Bhopal is that while the survivors mourn their dead, they ontologize not their dead but the perpetrators. If, as Derrida proposes in several texts (*Spectres of Marx*, 1994, notably) that mourning demands that we "ontologize remains, to make them present" (1994: 9), then Bhopal's survivors depart from this mode and ontologize Anderson, Chauhan, Mahindra and others. What is "made present" is the set of living people attributed with the responsibility for the disaster. This is uncanny because mourning here is for the victims as much as for the perpetrators who have never been caught. The faces of the perpetrators staring out at us from the effigies are a partial ontology of those who haunt the nightmares of the victims. The eerie uncanniness of the effigies' faces drives Bhopal's mourning not only in terms of the dead, but in terms of those *living* and responsible for the deaths. The topos and ontology of remains is with the living culprits as much as with the dead victims, suggests Bhopal. Just as the dead are kept alive through mourning, the Bhopal uncanny in the form of effigies, keeps the living perpetrator within the same framework. Nicholas Royle reminds us in his interpretation of Derrida that "there *are* phantom effects, even if phantoms do not exist" (278, emphasis in original). The Bhopal uncanny is constituted by the phantom effects of the unholy alliance—Indian politicians, industrialists, American experts and corporate heads. The phantom effects linger on, and the effigies are an attempt to ontologize these effects: the effigies are in fact embodied phantoms.

The effigies are phantom effects and embodied phantoms in the Bhopal uncanny also because, while everybody in India and the USA *know* the identity of the people in charge at UCIL and who are culpable, they remain elusive and unpunished by law. In this sense they are phantoms, ungoverned by any laws of the land (Indian or American), unavailable *materially* and *corporeally*. That is, these figures exist only as figures, as faces on the effigies, when it comes to the question of justice, although they exist in the flesh somewhere. This is a version of the uncanny *doubling* in Bhopal: people are available only as effigies, as constructed straw-men. This is a kind of phantomatic existence that is as far removed from their real, flesh-and-blood existence (Anderson in the USA, comfortably protected by his nation's laws and untroubled by any legal issues around UCIL, Chidambaram, Chauhan and other Indian politicians whose careers barely suffered a smear after the events). While their legacies continue to haunt Bhopalis (to which I shall return in the next chapter), they themselves acquire a spectral presence in the form of effigies.

The effigy is a marked departure from the traditional monster of horror where, as Noel Carroll has argued, the monsters are "impure and unclean . . . putrid or moldering things, or they hail from oozing places, or they are made of dead or rotting flesh . . . or are associated with vermin, disease" (1990: 23). Here the effigy depicts clear physical integrity—there is no decay or oozing collapse of the body. This is precisely, I argue, the monstrous of Bhopal's *in-corporated corporeality*. The entire monster is so well put together, so seamlessly integrated, from American and Indian organizations, lobbies, people and forms of state power that its full monstrosity is effected of unity rather than dissolution. What frightens us about Bhopal is the clear and unambiguous integration of the local (Indian) and the global (American), of Hindu scriptural/mythological figures and American symbols (the flag, the Dow logo). In sharp contrast, then, to the dissolving, leaky and ontologically uncertain body of traditional Gothic and horror, the Bhopal monstrous is embodied in the biomorphic *unity* of the effigy. Effigy burning is a symbolic destruction of Anderson, UCIL, Dow and the Indian state by the victims.

Photographs of the burning effigies literally disembowel and eviscerate the effigy. In a strategic reversal of the complete, unified and hybridized *body* of UCIL with its multiple in-corporations, the effigy burning reduces it to the bare minimum, literally taking apart the stitched together parts symbolic of the alliances, affiliations and liaisons that constitute the monster. If the traditional Gothic was constituted by the dissolving body, the Bhopal uncanny asserts itself in the form of the biomorphic coherent effigies that eventually lose their coherence. The monstrosity is in the unified nature of the UCIL body, and the evisceration, decades later, is the symbolic affirmation of this monstrosity that offers a minimal agency to the victims (survivors) of the monster. In the destruction of the corporate body of UCIL, in the form of the effigy-burning, we discern the collapse of the protectionism, the alliances and the evil miscegenated coalitions that made 2/3 December 1984. The collapsing body of the effigy is the tragically ironic tearing apart of the very processes and components that made UCIL one of the most potent symbols of modern India as well.

In the dissolution of the body, a process uncannily familiar to the Hindu funeral pyre, what burns is not a human, but rather an inhuman monster. The uncanny asserts itself therefore not just in the familiar (humanoid) nature of the effigy but in its dissolution as well. The monster which resembled a human form dissolves into a skeletal frame, also resembling a human form. Indeed, it could be argued that the uncanny is the startling effect of a monster reduced to the same skeletal frame as that of the human. We have other instances of such affiliations and in-corporations.

Other forms of liaisons and uncanny doublings are also visible in cultural texts on the disaster. Satinath Sarangi in an interview noted:

> "When we studied what medicines people were taking," he said, "we found that the chemical disaster had actually produced a windfall for the pharmaceutical companies—which are part of the chemical companies. So one part of the chemical companies poisons people and produces the market for the other part of the chemical companies. . . ." (Brown 2009)

Effigy burning, 1996. Courtesy International Campaign for Justice for Bhopal.

The editorial of *The Bhopal Marathon* writes without mincing words:

> You will learn how the victims, trusting that their government would come to their rescue, were instead let down by politicians who appear from the beginning to have been colluding with Union Carbide; how they were cheated by officials, bureaucrats and quacks; how when their water was found to be poisoned by chemicals leaking from Carbide's abandoned factory, they were denied clean water in defiance of a Supreme Court order and beaten when they dared to protest. . . . (5)

In *Bhopal: A Prayer for Rain*, Warren Anderson comments: "it is lucky we have good relations with the Indian government." Anderson is referring to the alterations to be made in the MIC storage sections to maximize productivity. The technical staff and even the American accountant tells him that the Danbury Headquarters (USA) would not approve of the changes, since safety is jeopardized. Later, we are shown scenes of open bribery and socializing between the Indian ministers (representing the state) and UCIL representatives.

Anderson's arrest and subsequent "escape" from Bhopal after 2/3 December, was, as was claimed by later commentators, effected by interventions from the very highest government offices in India (Gupta 2010). Investigations did not provide adequate answers to the motives in or processual flaws in setting up UCIL, the safety norms, legal jurisdiction and finally, compensation for victims. As the legal battle continued, it became obvious that UCIL was claiming, simultaneously, that the tragedy was caused by negligence by Union Carbide *India* Limited and so the parent Union Carbide company, which owned UCIL, was not subject to the jurisdiction of Indian courts and laws because it was *American*. The disaster was due to sabotage by UCIL's Indian workers, although the cut-backs on safety measures were decisions taken by Union Carbide USA. UCIL, it turns out, is both Indian and American, neither Indian nor American. The Asia-Pacific People's Environment Network (APPEN) *Report* states:

> Nothing illustrates the strange and weird collusion between administrators, scientists and doctors better than the question of cyanide. The manner in which officialdom, the establishment, clamped down in one way or another on any suggestion or clinical evidence that victims had died from cyanide poisoning, or those surviving may have continued to suffer from a form of chronic cyanide poisoning, is inexplicable. . . . (APPEN 1985: 4)

The report goes on to document how Prof. N. P. Mishra of Hamidia Hospital, who "put forth and defended, a theory of MIC damage that was limited to pulmonary dysfunction" had been "gifted 4 expensive sets of equipment, and a private ward consisting of two suites constructed by the company" (4). The statement by UCIL about this gift is available in Appendix 5.4 in the report, titled "Contribution of UCIL-APD Towards Community Welfare," which was a part of the "Submission to the Bhopal Poisonous Gas Leakage (1984) Enquiry Commission by CP Lal of Union Carbide." Once again we note the link between the biomedical and the neoliberal order where UCIL purchases biomedical secrecy through its funding of biomedical facilities—both with full consent of the government (Hamidia is a state-funded hospital). This link is the origin of the thanatopolitics that eventually produces the disaster.

Another aspect of the foreign-local collusion received a different response. Experts from the World Health Organization and doctors who came from all over the world to study the effects of MIC utilized the native case studies for their own research. N. R. Bhandari records that "private American doctors came at their cost for assessing the clinical condition, obtaining details of episode for publication of scientific material in their countries." This was facilitated by the locals: "some local doctors took out cornea (eye membrane) and sent the material to UK for analysis" (78). Later Bhandari notes how American lawyers arrived ostensibly to help the victims claim massive compensation: "They felt they would be able to claim millions of Rupees and get a big percentage from the victims" (104). Here the foreigner is a predator of an entirely different kind.[6]

The unholy liaisons standard to Gothic texts are visible here as well. The UCIL plant, processes and products are hybrids, as the cultural texts clearly indicate. Hybrids, writes Mary Douglas in *Purity and Danger* (2001), "are abominated." Kelly Hurley expanding on Douglas notes that hybrid abominations "shift from one . . . identity to another" (1994: 25). Yet, UCIL was not abominated for a very long time. Rather, it was hailed as a savior, a blessing and a source of prosperity. UCIL is Indian one moment, American the next, a curse and a savior, a local hero and a transnational messiah.

In many ways, UCIL was recognizably Indian: the relaxed, slightly indifferent working environment in the films suggest an Indian way of operations. The emphasis on safety (initially) and profits at every cost (later) is recognizably American. In one scene in *A Prayer for Rain*, Anderson touring the plant fixes a dripping pipe himself, and proudly declares, "my father was a plumber," thus reiterating the rags-to-riches but also the American entrepreneurial spirit in the Indian plant. This shifting identity between "Indian" and "American" produces the Bhopal uncanny for, to return to Hurley, "uncanniness is a product of the hybridization of the . . . form" (1994: 105). Even after the disaster, Satinath Sarangi writes, "the government, despite all that had happened still preferred to repose its faith in Carbide's tainted science" (*The Bhopal Marathon* 15).

The effigies also represent local versions of the devil. The uncanny generates, writes Hélène Cixous, a "mythic anthropology," a "foundation of gods and demons" (1976: 539). A certain primitivism, argued Freud, of the unformed is part of the uncanny (Freud 1971: 393–97). The return to Hindu mythology for images of demons in the protest campaigns asserts this dimension of the uncanny and the monstrous. We see this mythic anthropology developed fully in *Animal's People*.

> I feel it [terror] when I come back to this high place. I see Mother Kali stalking the forest below, her skin black as a roasted corpse. She's got these massive fangs and a red tongue hanging to her waist and a belt of chopped-off heads, each one wears a face of agony which is how they looked when they died . . . you see a black pipe climbing into the sky, I see Siva dark and naked, smeared

> with ashes from funeral pyres. His eyes are red from hash and smoke of burning flesh, dancing he's, from all sides I can hear the screams and cries of dying people. . . . Why did the Kampani choose this city to make its factory? Why this land? Is it by chance that the old name for this place is Kali's ground? Is it by chance that Siva her husband wears cobras round his neck? (32)

The boy merges Hinduism's most feared gods—since they are agents of destruction, mayhem and killing, Siva and Kali—with the company's site. He hears ghosts ("the factory is full of them, when a big wind blows, their souls fly shrieking up and down the empty pipes," 32) in the space. The mythic anthropology of the abandoned site is the space of modern ghosts, but also of ancient demons and gods of destruction from ancient mythology and scriptural texts. Sinha makes sure we note the parallels: the burnt corpses recall the mass graves and pyres from that night in 1984, the pipes wherein we hear the shrieks are those from which the gas shrieked into the air that night.

Such a mythic anthropology of UCIL also appears elsewhere in *Animal's People*. The night of the disaster, an old doctor tells Elli the American doctor who has come to serve in Khaufpur, "the moon was two-thirds full. It was shaped like a tear and as it appeared through the clouds of gas, it was the colour of blood" (153). Reading the cultural texts, one also finds hybridization of the ghost with mythic and religious figures from India's ancient texts. The factory itself is imaged as such a hybrid: "Like all the folk living round here, she's terrified that one night the factory will rise from the dead and come striding like a blood-dripping demon to snatch them off" (41). "Animal" himself is a human who is partially animal-like, because he walks on all fours instead of on two legs.

UCIL as monstrous is a hybridization of the form—of the factory, of working conditions, of processes and policies—as well as miscegenated socio-political origins. Being both inside and outside India, being local and global, the hybridized monster works to generate not only the uncanny but an autoimmune disaster.

THE MONSTROUS AS AUTOIMMUNE DISASTER

Reading 9/11 Jacques Derrida stimulatingly characterized it as an autoimmune disaster (2003). Derrida had written of the autoimmune sociopolitical condition of contemporary states and cultures elsewhere (1998, 2005). Derrida writes:

> What I call the autoimmune consists not only in harming or ruining oneself, indeed in destroying one's own protections, . . . committing suicide

> or threatening to do so, but, more seriously still, . . . in compromising the self, the *autos*—and thus ipseity. It consists not only in committing suicide but in compromising *sui-* or *self*-referentiality, the *self* or *sui-* of suicide itself. (2005: 45)

W. J. T. Mitchell expands Derrida's reading to discuss borders and boundaries:

> The stretching of the metaphor seems to be exactly the point. The limits, borders, boundaries of the body (politic), its relations of inside/outside, friend/enemy, native/alien, literal/figurative are exactly what is in question in the metaphor of the immune system. . . . (2007: 281)

This elaboration is a useful framework within which to read the Bhopal representations.

UCIL sets itself up as an Indian branch and a subsidiary of Union Carbide, the "American and transnational" industrial corporation, as Sarangi describes it. This installation is the opening up of Indian industry to the foreign corporate body. In Lapierre and Moro's *It Was Five Past Midnight in Bhopal* (see chapter 11), we are given a detailed tour of the processes through which this border was thrown open.

There is the erosion of sovereignty. Sovereignty, writes Michael Naas, is linked to the "auto," or "terms that define the self's or the subject's ability to return to and assert itself in its freedom" (20). In an autoimmune disorder, the very sense of self is compromised, eroded, when the organism turns against itself, the "very being or unity of self-identity" (Naas 2006: 23).

However—and this instantiates the Bhopal monstrous—humans, or nations, cannot survive without infusion from the outside either. Michael Naas writes:

> Autoimmunity is . . . a threat insofar as it compromises the immune system that protects the organism from external aggression, but as in the case of immuno-depressants, a chance for an organism to open itself up to and accept something that is not properly its own, to the transplanted organ, the graft, in a word, to the other. Without certain forces of autoimmunity, we would reject organs and others essential to "our" survival—whether we are talking about an individual body, a community, or a nation-state. Hence there can be no community without autoimmunity, no protection of the safe and sound without a perilous opening of borders. (25)

Derrida writes: "[autoimmunity is] a *double bind* of threat and chance, not alternatively or by turns promise and/or threat but threat *in* the promise itself" (2005: 82, emphasis in original). This dualism is the heart of autoimmunity.

Lapierre and Moro document the crisis in agriculture in India wherein the innovation of the Green Revolution produced a high demand for pesticides. When India could not produce enough pesticides, it turned to foreign manufacturers (63). This, it was hoped, would help ameliorate the "ancestral curse of its famines" (74).

The urgent need to ensure agricultural production within the country required, as Lapierre and Moro note, the entry of foreign manufacturers, of which one was UCIL. In other words, the necessity to ensure peace and prosperity, to protect and save the population and (from famine) within one's territorial borders demanded the entry of the foreign agent.

The entry of the foreigner was not a case of infection or contamination: they were invited in, and made comfortable within the sovereign territory of India. Lapierre and Moro write:

> According to municipal planning regulations, no industry likely to give off toxic emissions could be set up on a site where the prevailing wind might carry effluents onto nearby populated areas. . . . All things being equal the application [of Union Carbide] should have been turned down. But the Union Carbide envoys had taken good care not to mention that their proposed factory would be making its pesticides out of the most toxic gases in the whole of the chemical industry. (90)

Other factors were also instrumental in this prying open of Indian sovereignty, and Indian researchers and technicians were prepared to work for salaries ten or twelve times lower than standard American wages (90). The Indian politicians, bureaucrats and local bigwigs extended a warm welcome to the American Carbiders (101–02). The Carbide guest house at Shamla Hills had a suite "at the permanent disposal of the Chief Minister of Madhya Pradesh" (Lapierre and Moro 143).

In an earlier section I referred to the neo-imperial Gothic that is UCIL and its invasive, incorporated relationship with postcolonial India. Just as the imperial Gothic, as Patrick Brantlinger has shown, was characterized by very real fears of across-the-seas invasion by immigrants, refugees and aliens, the neo-imperial Gothic is marked by the invasion of Indian sovereignty by foreign companies, but with one crucial difference. This invasion is essentially effected through an in-corporation, of UCIL and other companies into India's mercantile, economic systems, manufacturing processes and even administrative mechanisms (for instance, the close links of the UCIL personnel with Indian government officials and political leaders). Thus, the neo-imperial Gothic that "produces" Bhopal 1984 is one that is initiated and processed through a carefully choreographed assimilation and in-corporation, an insidious invasion, if you will, of the foreign.

UCIL and its products are welcomed into India because they are seen as essential for its survival. That is, the borders needed to be kept open for the survival of the internal populace. Something "not properly its own" is brought in to protect its own. The immune system—such as municipal regulations and safety precautions—that were intended and designed to keep out foreign bodies did not function as they should, out of necessity or out of other concerns.

As cumulative effect then, the immune system of the sovereign state turned against the very (social) body it was designed to protect by welcoming the foreign object, UCIL. Admittedly, at the moment of inception, nobody quite knew whether UCIL was a threat or a great chance. H. L. Prajapati writes about the establishment of UCIL: "The Union Carbide had also made a huge breakthrough in the field of plant protection by setting up a huge research and development center at Shyamala Hills, Bhopal. . . . The center aimed at developing newer types of pesticides that suited the Indian conditions" (2003: 17).

Once the plant is up and running, Prajapati notes, "the plant authorities had not taken care to impart minimum knowledge to the population residing there about the precautionary steps to be taken in the event of gas leakage" (21). Now, autoimmunity, writes Michael Naas,

> stems not simply from the fact that we can never know whether we have chosen well or ill, whether something will turn out good or bad, whether it will have shown itself to be a threat *or* an opportunity, but, rather, from the fact that the opportunity *is* the threat, and the threat the chance. (28, emphasis in original)

India did not know till the first accidents in UCIL whether its decision to let Carbide in and make pesticides to save Indian agriculture was a good decision or a bad one. Its systems to ensure probity, safety and safeguard the interests of the population were shut down, or turned to ensuring profits for the (foreign-yet-Indian) company. UCIL was projected and welcomed as a chance. UCIL is:

> a guest who threatens to turn the host's immune system against itself, right up to and including the host's very capacity to receive or invite a guest, that is, right up to and including the sovereign power or ipseity of the host. The host thus needs the guest, or the parasite, to be himself, and yet the parasite also threatens not only the life but the very self-identity or ipseity of the host. (Naas 28–29)

Bhopal as an autoimmune disaster is the effect of this mixture of an invited chance metamorphosing into threat, of hospitality extended and the incoming foreigner providing the threat instead of the promised sustenance. UCIL turns the nation's sovereignty against itself in more ways than one.

The nation's guardians became UCIL's trusted lieutenants and supporters. When the disaster event actually occurs, if the stories have any credibility, Warren Anderson was allowed to return to the USA because of the intervention of the Indian government at local and national levels. The settlement arrived at by the Indian government did a disservice to its own citizens. Noted lawyer, Indira Jaisingh in her concluding essay to Chouhan's book writes:

> We must frankly acknowledge that the legal system failed the victims and then begin to consider reforms. . . . Given the New Economic policy which welcomes investments in every conceivable sector, we will see the emergence of multinationals in pursuit of cheap labor and markets. We need to introduce a system of laws which will make them accountable for standards of safety equal to those prevailing in their countries of origin. . . . Multinationals operating in India, must agree as a condition of doing business, that they will submit to the jurisdiction of the Indian Courts. (Jaisingh 1994: 211–12)

Jaisingh is in fact calling for a defense of Indian sovereignty here, wherein the business corporations that enter this sovereign state cannot be allowed to destroy the immune (for instance the legal) system that safeguards its citizens. Jaisingh emphasizes the conditions under which a foreign body may be allowed to enter and operate in India. She draws attention to the necessity of sovereignty in any discussion of foreign capital. Jaisingh suggests that we learn from Bhopal: "Why have the Indian authorities not woken up, even after Bhopal, to amend our laws?" (212). This failure of the Indian legal, medical and regulatory mechanisms that enabled UCIL to function the way it did, leading to the disaster, is summarized in an eyewitness account. Lalit Shastri notes that "the safety of the citizens of Bhopal was overlooked in preference to the interests of the multinational corporation" (78–79). This was the erosion of the sovereign authority of the nation.

"Sovereign authority," writes Jacques Derrida, "is having the power to decide, to be decisive, to prevail, to have reason over or win out over . . . and to give force of law" (2005: 13). It is a form of self-determination and selfhood, he argues. It is indivisible and unitary. Such a sovereignty closes itself off from the outside, from the Other. If sovereignty is aligned with autoimmunity, as a closing off of borders and turning to oneself, then such an "autoimmunity compromises not the life of some identifiable *autos* but the *autos* itself as the foundation and guarantor of identity" (Naas 24). If we conjoin Jaisingh's argument with that of Derrida's then we have an interesting definition of sovereignty itself.

One cannot be absolutely immune or sovereign, since, as was clear in the years leading up to the arrival of Union Carbide, India needed outside help to cope with her internal demands—agricultural, medical, military, among others. An eyewitness, Lalit Shastri puts it thus: "for the survival of crops from these hybrid seeds it became essential to use pesticides and fertilizers" (74). Shastri notes that "the role of the government of Madhya Pradesh had also been very crucial" (1985: 76). Decisions about UCIL, as these texts clearly indicate, were taken in order to safeguard Indian agricultural production. Once India had embarked on the Green Revolution, it entailed changes in agricultural practices that fuelled the demand for pesticides and fertilisers. Given India's inadequate production of the latter, the two options were: to either let the Green Revolution falter or to allow such manufacturers. The latter arrive, therefore, as a direct result of the state of affairs in—and discussions around—India's agricultural and industrial sectors.

A sovereignty that closes itself off can no longer survive. That is, sovereignty cannot be entirely self-identified and self-determined. Hence, Jaisingh proposes an assertion of sovereignty that *acknowledges* the necessity of the Other arriving in India. Rather, she calls for learning a lesson from Bhopal. If Indian autoimmunity caused Bhopal then, as is the case with the immune system (which, along with the nervous system, is one of the two bodily systems capable of learning), we need to learn from it and strengthen the immune system, modify it.[7] W. J. T. Mitchell writes:

> The nervous system can accelerate its learning process with self-conscious reflection, critique, the preservation of memory and history. Immunity is a form of cellular memory; the body learns by experience how to fight measles, and it doesn't forget. The most dangerous threat to the immune system, then, is amnesia, the forgetting of what it has learned. . . . (284)

The immune system remembers the attack, the contamination, the virus, and recalibrates itself for the future. Thus, the strengthening of the immune system demands the contamination. The sovereignty of the body, to be protected by the immune system, demands that this system be prepared, guarding the borders, precisely by inviting the foreign in. In other words, sovereignty demands a welcoming of the foreign, the Other, the outsider.

Bhopal, like India as a whole, needs to therefore keep memories of the tragedy alive so as to immunize itself better by way of better laws and a greater commitment to its citizens. Remembering Bhopal—and this returns us to the uncanny, of the ghostly survival of the past—should be instrumentalized, says Jaisingh, to ensure that we learn from it, and activate the alteration of the immune (legal) system so that we do not suffer autoimmune disorders.

In the Gothic "even triumphs of the built environment are precarious" (Wasson 132). Here the triumph that the UCIL first represented and symbolized enables—indeed generates—the culture of precarity for those dwelling within and without. In the slow violence of Gothicized Bhopal, the built environment literally and metaphorically seeps into the natural environment, poisoning all.

With the dubious alliances forged between UCIL and the local authorities, the state that functions as the immune system turned against its own organs, and this is precisely the destruction of the country's sovereignty. It is this sort of invasion of the immune system that ought to strengthen it further, which is precisely what Jaisingh's plea (for greater accountability laws of MNCs operating in India) implied.

NOTES

1. The APPEN document refers to one Dr Ragu, while Bhandari refers to a Nagu, although both are described as "Director, Medical Services, Bhopal" in these texts.

2. The Minutes of the Review Meeting of the Indian Council of Medical Research, Researches on Health Effects of Exposure to Toxic Gas at Bhopal held on 3–4 May 1985 at Valladh Bhavan, Bhopal, has an interesting moment recorded. The chairman at the meeting "informed about the desirability of bringing out scientific publication and communicate the findings. He suggested that if so desired the *Indian Journal of Medical Research* will bring out a series of special volumes to report the findings of the studies. . . . One member cautioned about the legal implications of the publication. The consensus, however, was that the findings should be published as early as possible" (APPEN, appendix 5.16, 142). N. R. Bhandari notes, "in spite of all efforts the Government did not give permission for publication of most of the papers [from the studies of Bhopal victims]" (2010: 73). The emphasis on information is central to all cultural texts on Bhopal.

3. The unnameable "Thing," argues Gary Farnell, is the unspeakable, the unrepresentable, which is both inside language/culture and yet outside it, and generates the Gothic effect (2009).

4. The Chidambaram effigy was burnt again in 2012 in protest against the decision of the Indian government's Group of Ministers on Bhopal to not revise the figures of deaths and injuries caused by the disaster in the curative petition pending before the Supreme Court.

5. Emma McEvoy refers to " the tradition of Gothic fragmentation" (2007: 24).

6. N. R. Bhandari records that "within a few days government (Health Department) issued an order that these foreigners should not be allowed to examine the patients and take their case histories" (78).

7. The immune system learns by clonal selection: it produces antibodies that mirror the invading antigens and bond with them, eventually killing them.

Chapter Four

Bhopal's Biopolitical Uncanny II

The Haunting of Nature

Extending the study of the biopolitical uncanny in these texts, the present chapter returns to bodies and landscapes in the aftermath of the tragedy. It is particularly interested in the construction of landscapes of memory around the dead and the disappeared, the maimed and the surviving.

The Gothicization of Bhopal involves the generation of a biopolitical uncanny which ruptures the boundaries of bodies—material-corporeal, communitarian-social and corporate. The precarity that this engenders due to the collapse of frames of recognition and consequently of cultural protection is not merely restricted to the events of 2/3 Dec. 1984. Both slow violence and horrorism, in the Bhopal Gothic, is about the incremental collapse of these bodies, and of the Nature itself upon which the bodies depend and in which they are embedded.

The biopolitical uncanny begins, expectedly, with the body. From the body it moves to the state of the immediate environs, the family, and the effects of the disaster in the family. From the haunted families of the texts to the extensive haunted community and the city, the chapter opens out its study of the slow violence and lingering toxicity. Finally, it argues that Bhopal's texts document a haunted body politic. Haunting is a thanatopolitical theme, and not merely a psychologism in the cultural texts of the Bhopal Gothic.

BIOLOGICAL CITIZENSHIP AND ITS UNCANNY

The survivors of the tragedy possess only one kind of citizenship, that of the biological. Adriana Petryna's definition of "biological citizenship" already cited but worth repetition goes: "a massive demand for but selective access to a form of social welfare based on medical, scientific, and legal criteria that both acknowledge biological injury and compensate for it" (2002: 6). That is, "the damaged biology of a population has become the grounds for social membership and the basis for staking citizenship claims" (5).

In the immediate aftermath of the events of 2/3 December 1984, most of the living bodies became "foreign" to the norm of the human (what I have previously described as the abject). Over the years several of these "bodies" battled for survival and biological citizenship because they harbored something foreign within them. Their biologies ineluctably altered forever after the body inhaled MIC, their only identity was, one would assume, that of the "gas victim." However, as Petryna has noted in the case of Chernobyl's victims, even this biological citizenship based on their injuries was not easily acquired, and in many cases outright denied. Here is an account of one of the Bhopal survivors printed in the newsletter *777*:

> By 1996 Idris was coughing badly. He felt very ill, and developed a fever that would not abate. . . . Idris was given a cardiac test, and discovered to have an atrial septum defect (a hole in the heart). This was the root of his problem, but his condition had been made infinitely worse by his exposure to Carbide's gases. His lungs are severely scarred and by now have lost half their function. . . . THE HEART SURGERY IDRIS NEEDS IS AVAILABLE at the Bhopal Memorial Hospital. It was established (with funds raised by selling Union Carbide's Indian assets) supposedly for the benefit of gas victims, but Idris has been told that he must pay for the operation. He does not qualify for free care, say the hospital's administrators, *because he is not a gas victim*. But of course Idris is a gas victim. There are dozens of witnesses to attest to the fact and Sambhavna has written to confirm that Idris's deterioration is gas-related. Sorry, say the hospital bosses, if he is a gas victim, where are the documents to prove it? Vainly, we point out that the papers in question have been ashes these fourteen years. Forcing the authorities to issue new papers will take court action, maybe lasting years and time is running out. (*777*, May 2006, emphasis in original)

In another example from the newsletter, we are shown a little girl showering under a street tap. The caption informs us:

> This little girl Manakshi may look like a three year old but she is in fact seven. Manakshi suffers from a growth disorder and has been undergoing treatment. . . . (*777*, Summer 2011)

Other such visual records in *777*, the Bhopal Medical Appeal newsletter (2016), document numerous such disproportionately formed, "anomalous" bodies. These cultural texts are marked by a representation of the biopolitical uncanny that haunts Bhopal.

The Gothicization of Bhopal's bodies and Nature ensures through a process of in-corporation the anomalous bodies of Bhopalis. There is no "outside" horror any more, since the bodies have been othered to themselves through the poisons ingested. William Hughes has argued in the case of "Gothic tourism" texts that when protagonists enter English or European wilderness (the poorer areas of the cities) only to discover horror they in fact stumble across the "a savage, atavistic past within which the self rapidly becomes implicated" (Hughes 2003: 123). That is, there is no external source of this horror: the horror is the decayed, decadent, deteriorating horror of the home-grown variety, local, indigenous and regional. While the Gothicization of Bhopal did involve the foreign body, as argued earlier, the native body is now host to the foreign. Just as the Nature or environment of Bhopal incorporates the foreign, the Bhopali body is a host to the foreign chemical body. The self is implicated in the "atavistic past" dating back to 1984: atavistic because what is iterated in the present-day body is the reiteration of that day. To phrase it differently, the present-day Bhopali body, whether human or Nature's, is an atavistic reiteration of events, processes and chemical *bodies* (MIC) from thirty years ago. The Bhopali self is no longer *Bhopali*, but implicated in a past whose origins, contours and content are the result of the then incorporations of the foreign.

The uncanny results from the experience of one's own body *as* foreign, or of a foreign body within oneself (Royle 2003: 2). In the case of Bhopal, we see such an uncanny emerging in the stories of the survivors.

In Idris's case there is a misrecognition and denial of biological citizenship by the institutions whose responsibility, in the wake of the disaster, was to ensure this very citizenship. Idris does not have a body that fits the norm, and neither does Manakshi (also published in *777*). While the former's biological citizenship has been denied him, in the latter what is at stake is a cultural citizenship—being of a certain biological age in her cultural settings—because her biology defeats any cultural recognition.

Bodies such as that of the Bhopal survivors are foreign in the sense that not only do they *not* fit the accepted cultural norms of the human form/body, they are also denied a biological citizenship founded on their anomalies and aberrations. In other words, the survivors experience themselves as foreign because their bodies are foreign to the norms and to the institutional/cultural frameworks that enable citizenship.

I further propose that the biopolitical uncanny is the cumulative effect of rendering Bhopalis foreign to themselves exactly in the same fashion as UCIL and later Dow's categorization of the plant as foreign to Indian law (Anderson refusing to stand trial in India, UCIL declaring itself outside Indian juridical jurisdiction). That is, the body of the plant, the *things* from UCIL, and the survivor are all rendered foreign to themselves due to the complete denial of their subjectivity, form and experiences. The individuals experience bodily uncanniness because, as Andrew Warsop writing about ill bodies and the uncanny argues: "[in] bodily uncanniness, we become aware of how our bodies are like other objects not merely in the sense that we see them as perceivable and outside us but other in the sense of being inert or life less material" (2011: 493).

This perception of oneself as lifeless and inert is stated unambiguously by a survivor, who refers to the victims as "zinda laash" (living corpse) in at least one documentary. Another survivor, who lost her four-year old son that night, says in an interview to Suketu Mehta:

> We are sobbing, but death doesn't come. Like a living, walking corpse you have left us. . . . If with the strength of your money you had shot us all at once with bullets, then we wouldn't have to die such miserable sobbing deaths. (Hanna et al. 2005: 115)

Living death is abjection (Farnell), and the survivors are at once inside and outside themselves, uncanny. They are "border entities" (Hurley) characteristic of the Gothic, between life and death, and in them *embody* the Gothicization of Bhopal.

Appeals and campaigns took recourse to the language of abjection to highlight the erosion of normative and biological citizenship. In fact, the discourse of protest employed the term "genocide" to describe the UCIL and state attitude toward, first, the Bhopalis and second, the victims. Here is APPEN's pamphlet:

> Clearly Union Carbide viewed Indian life as cheap and expendable, an outlook which is the basis of genocide. In their arrogance, criminal neglect and misinformation were seen as responsible behaviour. The Bhopal gas victims were seen as only guinea pigs and "doctors" from abroad rushed in to measure and document the effects of MIC poisoning. . . . The Bhopal genocide raises fundamental questions concerning the nature of the development process and the role of Transnational Corporations in the Third World countries. . . . (APPEN, Appendix 5.26, 1985: 174)

Accounts of the diseased body, like anatomical drawings, are reflective reconstructions of the physical body, an externalization that is far removed from the actual body, but is somehow its uncanny double (Kenny 2012). The Bhopal victims' dead bodies are reconstructed in verbal autopsies through conversations with their near and dear and acquaintances, states a report in *777* (Oct. 2003). The verbal autopsy, as defined by the Centre for Global Health Research, is the collection of "written narratives from families or other reliable informants in the local language describing the events that preceded the death. In addition, answers to standard questions about key symptoms were also recorded" (http://www.cghr.org/projects/million-death-study-project/what-is-verbal-autopsy/. Verbal autopsy standards as issued by the WHO are to be found at http://www.who.int/healthinfo/statistics/verbalautopsystandards/en/. 8 Sept. 2016). The "recovery" of diseased bodies, so to speak, in the form of the verbal autopsy report is at once an instance of the biopolitical uncanny, following Steven Kenny's formulation on the diagrammatically reconstructed human form, and an assertion of the biological citizenship of the dead. They double up for, work as substitutes for, the dead, even as they point to those institutional mechanisms that deny the dead and the living full biological citizenship. While the dead are reanimated, the living survive as the living dead.

There is another sense in which the biopolitical uncanny might be interpreted in the case of Bhopal's surviving "bodies." As noted earlier, the Bhopal uncanny stems from a lack of frames of recognition, itself stemming from secrecy around crucial information and data. The Bhopali survivors experience their bodies as foreign to themselves because there is an unknowable/unknown secret within their bodies. The foreign object—phosgene, cyanide, MIC—is lodged deep within their DNA in many cases. The "foreign," as is characteristic of the uncanny is within and thus renders the body familiar yet unrecognizable. Therefore, it is possible to argue that the secret encrypted in UCIL chooses the bodiesof Bhopal victims as the site of its inscription.

If the DNA is the code/book of life, then it is a language. The chemical formula of the gas inhaled is, likewise, a language. Due to the structures of secrecy around the formula, the constituents and the effects that the text of the gas had upon the book of life, the inscription (poison gas) does not appear clearly, legibly. Rather, it appears in the twisted, suffering bodies of people, in the distorted physiognomies and limbs. Interpreting the bodies of Gothic characters, Christine Berthin writes:

> A character is a set of "characters," that is a set of signs in a determined system. Characterization is reduced to the inscription of legible signs on the skin of paper heroes. The letter of inscribed traits of personality has taken over the spirit of characterization, and the notion of private self has given way to mere markings on the visible surface of the being. (2010: 74)

Expanding this line of argument, I propose that the biopolitical uncanny of Bhopal manifests in the legible signs of the individual bodies, whose private, suffering selves have been erased in favor of their surface meaning as "gas victims." The inscribed traits of personality stem from the gas and its effects, the secret inscribed into their bodies alters what they have been as human individuals. This itself is a horrific Gothicization: an element (chemical) *not*-meant to be inside the human body has intruded and sedimented within it. The breaking of the border between inside and outside effected by the gas molecules causes a breakdown of the border between living and nonliving.

Now, unlike twisted and visibly de-formed and dehumanized survivors, there are also bodies whose form does not reveal what they carry inside them. Invisible monstrosity of diseased bodies, argue critics, are also instantiations of the Gothic (see Fahey 2015). In the case of inherited genetic anomalies, such as those "embedded" within the bodies of Bhopal's survivors, the monstrosity is hidden, invisible but no less powerful and hideous than the twisted, misshapen one. The Gothicity of these bodies lies precisely in the invisible monstrous, but a transmissible monstrous.

While it seems inappropriate to treat twisted, suffering bodies as semiotic texts of the gas, what is irrefutable is that the foreign is present *as*, is embodied as, haunted, material bodies. The events of 2/3 December 1984 linger on in the skin, physiologies and anatomy of the individual survivors. The body is a crypt into which the secret has been inserted, or the body is rendered a crypt due to secreting of the secret poison. The biopolitical uncanny is, then, the haunted, rendered-crypt of the individual, whose individual personality or character is subsumed under the tag "gas victim," whose material body becomes the expression of a secret text within. David Punter examining the crypts of Gothic texts says:

> "Gothic" would signify a field of hauntings, of course, but also a dealing with the issue of namelessness, a field in which, as in the first words of *Hamlet*, the very notion of stable naming is thrown into doubt, as, typically perhaps, in the convoluted family trees of *Wuthering Heights*, which are, undeniably, unintelligible, linked only by repetition—repetition of names, repetition of syllables—in a mad dance of instability within which all notion of lineage is unseated according to the "law of the orphan," so that the desire for, and the terror of, anonymity. . . . (2012: 256)

"Natural" bodies in Bhopal are now haunted, anonymized as "victims," reduced to a number on a datasheet, on legal documents. There is no stable naming and, given the large numbers of missing, disappeared and unidentified dead from 1984, such a naming or numbering is not even possible. The namelessness is not only a feature of the dead and the disappeared: in many ways it is also a haunting feature of the survivors. Indeed, it could be argued that what is haunting about Bhopal is the cryptonymy: the subsidence, subsuming and secreting of individual identity within and under the name "gas victim." UCIL's processes therefore dehumanized the individual by erasing all identities except this one, of victim. If in the case of the dead the human name is inscribed upon stone and memorials, the Bhopal biopolitical uncanny ensures that such a name cannot be inscribed anywhere, although the bodies themselves are sites of inscription—of the gas. The "achievement of anonymity . . . is akin to the defamiliarizing of the uncanny," writes Punter (258), and this is precisely the biopolitical uncanny of Bhopal.

HAUNTED DOMESTICITY

From the abjected, monstrous and uncanny bodies of survivors, I now shift focus on to another kind of haunting that is discernible in Bhopal, and that is the haunting of families, the domestic scene and the community itself.

Francesca Moore's *Bhopal: Facing 30* is a collection of staged studio-shot, family photographs. The aim, she writes, was to "reference the traditional Indian studio portraits usually acquired by higher castes . . . [that] demonstrates wealth, accomplishment, and a sense of achievement." The Bhopal survivors, she believes, "could register their own dignity, values and resilience through the medium of the family portrait" (2014: 9). She continues: "I photographed the families with the belief that they are as good as anyone else" (13). The captions accompanying the photographs are interesting. They list the names of the individuals in a family, their age and gender, current place of residence and their place of residence in 1984. The "notes" accompanying the caption usually have "gas affected" as an entry. "Gas affected" as a descriptor functions as a family name, so to speak, that unites the entire community.

Marianne Hirsch in a famous argument has proposed that family portraits generate an "affiliative look" and a "familial gaze" (2009). She argues that viewers identify with the family portrait because we are responding to "dominant mythologies of family life" even as the family members in the portrait/photograph "define themselves in relation to each other in the roles they occupy as mother, father, daughter, son, husband, or lover" (vi). Such portraits, she proposes, are situated, and can be read at, the intersection of the public and the private, "between the family album and the public memorial, between personal and political expression and meaning" (vii).

Moore's staged studio portraits of Bhopalis and the survivors of 1984 is a remarkable collection, and generates a haunting and an uncanny. Of the family is "both place of safety and potential site of terror" (McGuire and Buchbinder 2010: 298). The families in Moore, we realize, are haunted by the terrors of the events of 1984, and the sense of loss.

Moore foregrounds the familial and the affiliative in the portraits. Thus, she appears to reinforce the dominant mythologies of family life, as Hirsch has argued. She calls upon us to respond, via the familial gaze, to the portrait of the "Indian family." Yet, where this dominant mythology comes unraveled is the caption and references in the "notes" to the families' place of residence in December 1984. In some cases, the notes also inform us of a dead family member, but without stating cause of death. Patterns of domesticity are at once reinforced and undermined when we are told of the poisoning within the family. We begin to see the individuals, with their names and identities as related to each other not only as related to each other in terms of the roles they play (Hirsch) but also in terms of being victims. In some cases, the term "gas affected" is applied to the entire family, without identifying any particular member—for instance, with Zaheer Ansari and family or Mohd Khan and family (unpaginated). In other cases, we are told that one of them is "gas affected": for instance, in the family portrait of Nasir, Afsana and their four children, Nasir alone is identified as "gas affected." The poison gas is the disruption of the home and family. Tamara Wagner (2010) has argued in the case of the Victorian age's "clinical Gothic" that substance abuse and poisons were presented in sensational novels as disruptive of patterns of domesticity. Likewise, Moore's family portraits alert us to the invisible, insidious and *material* substance—MIC—that "frames" the family, even though this frame is within them. (I shall return to the theme of *material-* substances in chapter 5.) It generates Gothic-like patterns of disrupted domesticity when we recognize families as connected not only through the roles they play but by the poison they ingested and which frames and binds them. Families, one could then argue, are haunted by the gas that still envelops them and the family. In other words, Moore forces us to see the family portraits not merely as people who retain their dignity (as she stated in her Introduction to the volume) but who endure their suffering, their losses and their mourning with dignity. Thus, the formal setting of the family portrait in Moore's collection is about dignity that simultaneously inscribes a visual ellipsis (I adapt the term from Terence Wright, 2008) around and within the family: the poisoning and the suffering. The portrait is built around such a visual ellipsis, of a suffering that finds, literally, no place in the photograph and yet exists as a haunting.

There is nothing in the portraits themselves to indicate what or how the families and individuals have suffered. There is nothing that connects the people to Bhopal, 2/3 December 1984. To return to an earlier argument about family portraits from Marianne Hirsch:

> It is precisely the displacement of the bodies depicted in the pictures of horror from their domestic settings, and their disfiguration, that brings home (as it were) the enormity of Holocaust destruction. And it is precisely the utter conventionality and generality of the domestic family picture that makes it impossible for us to comprehend how the person in the picture was, or could have been, exterminated. (1992–1993: 7)

Replacing "Holocaust" with "Bhopal" we find that Hirsch's reading of the ostensible "normalcy" of the family portrait enables us to see the uncanny in operation. MIC, gas, poison and death hover as the context, if not the content, of the portraits. Hirsch in her reading of Spiegelman's *Maus* claims that the photos in *Maus* stand in for what does not exist and for "what has been" and what is lost (9).[1]

The family is haunted, framed and limited by the history of the gas inhalation. The gas and the poison are the secret the portrait does not reveal, but is at the heart of the family, and is indeed the reason *why* Moore took the photographs in the first instance. In the staged dignity of the studio portrait the visual ellipsis is what haunts us viewers.

The family portrait of the Bhopal victims is haunting in another sense. Its publicness is at odds with the suffering we know is at the heart of their personal stories. Thus, the family picture quietly frames the personal suffering and story as a part of the project of public memorialization. Moore stages the family again, and simultaneously *decontextualizes* the family from 2/3 December 1984 and *recontextualizes* them within public memory. The survivor families are presented in the present. Their memories of the past are not given to us (and are thus decontextualized from December 1984), but the very presentation serves as a re-presentation of the family from the past *as* gas-affected from the past where it is their connection to the fateful events that generates today's presentation. Personal memories slide into public memory because of this re-presentation.

The family portrait is an instance of "post-memory" (Hirsch 1992–1993). Post-memory "is distinguished from memory by generational distance and from history by deep personal connection" (8). The adults possess the memory of 1984, which excludes the children. The children *know* 1984 through their family's history. Photography brings memory and post-memory together. Within the frame of the photograph children migrate into the personal memory of their parents, they claim it for themselves (as post-memory).

The survivors are human documents in a documentary. Moore in fact brings the survivors together as a human document. The documentary as a form is given to the presentation of evidence (Chute 2016: 2), of recording and archiving (5). Documentation—derived from "docer," meaning "to teach" and "to show," and which lies at the heart of "documentary"—is an epistemic project, "the kind of knowing that is all wrapped up with showing, and showing wrapped with knowing" (Gitelman 2014: 2). Moore shows us a family wherein some members *know* the events of 2/3 December 1984, a "'knowing" into which children are drawn in via post-memory, and into which we are drawn in as the audience/readers. One could go further and argue that these family portraits, with the palpable visual ellipsis at the heart of the representation of domesticity, are more than documentaries: they are memorials. Marita Sturken in an essay on photographs of Japanese internees in the USA during the Second World War has argued that the "most powerful kinds of memorials demand forms of reenactment in the sense that they force viewers to participate rather than to find a comfortable distance" (1997: 703). Moore's photographs stage the domestic scene, the family and the familial relations for us, but in the process they reenact a traumatic participation: we do not see these as the "routine" family photographs that capture, as Moore put it, aspirations, wealth and achievements. We participate by *recognizing* the visual ellipses, the poisoned "frames" that define the family's aspirations and achievements and the poison-determined roles of the family members. That is, we recognize more than the roles of mother, father or husband: we recognize the "gas affected" individual whose role in the system of relations is also determined by the poison inside her/him.

The family-under-siege is a familiar genre in Hollywood and horror films, as well as in the traditional literary Gothic. Commentators such as Tony Williams have observed that the family is the cause (and site) of horrific events (1996). Moore's Bhopal photographs gesture at the family-that-could-have-been, but for the horrific interruption of 1984. The postcolonial nation, often metonymically symbolized in its literature in the form of the bourgeois family (Schultheis 2004), is, then, disrupted when the family is disrupted. To draw another parallel, in traditional English literature, the Gothic was a genre wherein "a Gothic home offered a strategy either to reaffirm or to puncture the idea of a domestic shelter from all threats [of] . . . new technologies, the results of rampant industrialization—and to new awareness of mental illness, crime and the anonymity of expanding urban spaces" (Wagner 2014: 110). Yet, "the narrative as a whole harnesses Gothic terror within a homely setting" (111). For Bhopalis, many of whom are searching for their missing family members and wherein survivors are haunted by memories of the missing *and* suffer from assorted illnesses, the family is where immediate terror is *embodied.* Moore's family portraits may be seen as implying the *Gothicization of the family* and the domestic space itself where disease (including that staple fare of domestic Gothic—mental illness) disappearance, decay and death define family relations. The missing and the injured haunt the empty spaces within the home. The post-disaster imaginative recasting of the family highlights not the unity or belonging but the MIC-driven fractures. The portraits might also be read as expressing a desire, in its staged reconstruction of the family as a unit, to return to a past when the family was entire, and family history was not a toxic history.

In one sense, then, the photographs are those of *haunted* men and women: haunted by memories, and by something inside them that will not (let) go. Their identities, despite Moore's efforts, are woven around the crypt inside them, inside their families and inside their homes. Thus, while conceding that the family photograph bestows a certain dignity—of survival—it can also be argued, as I do above, that it enacts a haunting by staging the family with the poison inside them, a domesticity built around not just family roles and relations but around loss, suffering, pain and death.

Gothicization, then, is haunting by any and all means: the "unnatural" chemicals inside the human that renders the human nearly dead or barely alive, the memories of the dead and the missing in the minds of individual survivors and in the collective memories of the family, and the haunting loss of entire communities.

HAUNTED COMMUNITY, SPECTRAL CITY

If the family photograph alerts us to the visual ellipsis at the heart of the family, the city of Bhopal is, by *nature*, now a "spectral city" in the cultural representations. I take the term from Julia Round who argues that a spectral city is one "that, despite being buried in history, nonetheless still exists in the public consciousness." Her example is that of the Ripper legend which is determinedly kept alive in collective memory and the city of London itself is marked by a "sense of haunting" (2012: 339). The haunting is due to an identifiable crypt in Bhopal. At the heart of Bhopal is not only UCIL—an abandoned ruin, a grave—but also the memories of mass burials/cremations and large-scale topos of death. All documentaries on Bhopal take us on a journey through this crypt of a factory. In Indra Sinha's *Animal's People* it is described thus:

> Step through one of these holes [in the factory wall], you're into another world. Gone are the city noises, horns of trucks and autos. . . . No bird song. No hoppers in the grass. No bee hum. Insects can't survive here. . . . (2007: 30)

A crypt is the place from which the ghost/s emerge/s. The crypt can "take place" by producing concealment (Castricano 2001: 27). It is built through violence and a penetration of the crypt demands extreme devices. But the crypt is also at the heart of the Gothic, and is marked by its ability "to isolate, to protect, to shelter . . . from penetration," as Eve Kosofsky Sedgwick has argued (quoted in Castricano 27). The crypt is incorporated into the city (27–28). It is a "foreign place, prohibited, excluded within the 'Self' where it implies the topography of an *other*" (128, emphasis in original). Sinha continues the above cited description thus:

> Imagine you're in the factory with me. See that thing rising above the trees, those rusty pipes and metal stairs going nowhere? That's the place where they made the poisons. . . . Each big wind pulls more iron sheets loose. We hear them banging like angry ghosts. All that's left now is its skeleton. . . . (30)

Contrary to Round and Castricano's interpretation of the crypt, I propose that the events of 2/3 Dec. 1984, itself produced through the encryption of necessary knowledge (as discussed in the preceding chapter), render Bhopal a crypt. As noted in the earlier chapter, the mass graves render the entire city a landscape of death, a geography of helplessness. But this interpretation does not adequately account for two additional features of Bhopal's tragedy: disappearances and anonymity, and continuing pollution. The crypt-that-is-Bhopal is produced as an act of violence—of 2/3 December 1984.

Photographs of those who disappeared on that fateful night are regularly exhibited at the annual protests, and are now a part of the Bhopal textual canon, so to speak. Yet UCIL's statistics deny the numbers of the dead or the disappeared.

The disappeared and the unidentified body is "a paradigm of the past itself, which is both continuous with the present and discontinuous from it; which simultaneously is and is not" (Domanska 2005: 404). The missing represents a "manifest absence" (404) which signals a "non-absent past" (404–5). Domanska continues:

> The non-absent past is the ambivalent and liminal space of "the uncanny"; it is a past which haunts like a phantom and therefore cannot be controlled or subject to a finite interpretation. It is occupied by "uncanny ghostly artifacts," which undermine our sense of the familiar and threaten our sense of safety. (405)

Bhopal's haunting is the space of the empty grave, the missing dead body and the inability of several of the Bhopalis to mourn their dead. It generates a space where "desperate human desires to preserve the delusive continuity of life (history, memory, monuments—immortality) do not reach but where the continuity of death manifests itself. . . ." (406).

Domanska's argument enables us to see the making of a spectral city and the haunting of urban nature in Bhopal. The city is identified globally *in the name of the unnamed.* It is a spectral city because its identity also relies extensively upon those who were *not* identified and have remained unidentified from 1984. Bhopal is rendered at once familiar and strange due to the staying-alive of manifest absences, of the dead. Raghu Rai snapped a man displaying a large sheet of over a hundred photographs of missing persons: they are dead but present as reminders of the events. The photographs reanimate the dead, ensuring that they stay on (Rai, *Exposure*, www.magnumphotos.com).

This non-absent past is rendered into a manifest absence through the active circuits of justice and memory campaigns. The haunting of Bhopal's urban space—its nature, so to speak—is initiated and reinforced by the issue network around the dead and disappeared individuals. Issue networks constitute a form of the public sphere in which an issue "sparks a public into being" (Renshaw 2013: 39). Renshaw continues:

> The most active publics are sparked into being by issues that are complex, with implications that are far-reaching and distributed beyond the concerns of any single individual or constituency, particularly issues that are novel and evolving. . . . (40)

Families of the dead/disappeared clearly have a stake in highlighting the identities of those dead/disappeared. However, Bhopal's justice and memory campaigns brought together people with no connections to the dead: that is, the issue of the unidentified dead/disappeared pulls together a melancholic public sphere. It is not specific to any family or individual, but concerns Bhopal as a community. Campaigners in Brighton, Durham and across the world, as members of the International Campaign for Justice for Bhopal, also keep these memories alive. This melancholic public sphere is transgenerational since we see people who were not born or had been infants in 1984 also being/becoming a part of the public, but it is also constituted by multidirectional memory wherein campaigners from across the world converge their efforts at obtaining justice for Bhopal.

Landscapes of disappearance, Minoli Salgado writes, embody a certain kind of reality: "reality is marked by the bodies of the disappeared—a landscape, it could be said, marked by the truth of disappearance and the disappearance of truth" (2013: 212). The photographs and family memories of the disappeared construct a landscape of disappearance denied by UCIL. This tension between an unofficial, vernacular public memory of the survivors or the annual protests built around this memory and the UCIL/Dow/Indian government's continued denial and silence over the causes, effects and lingering consequences produces the crypt that is Bhopal, a spectral city where public memory keeps 1984 alive, but those tasked with protecting the citizens do not.

The spectral city is a city with a crypt. The missing bodies, the lists of names and the photographs bring back the disappeared and the anonymous, animate them as testimony and constitute the city's identity. Following the work of Achille Mbembe, Suvendrini Perera writes:

> The public meaning of these bodies inscribed by necropolitics in "the maximal economy of the massacre" as nameless, undifferentiated deposits, nonetheless resides in their being made present, in the register of anatomo-politics, as missing and disappeared fragments, in order to testify to the very reality of necropower. . . . (2006: 649)

Perera is gesturing at a narrativization that brings back the dead and the disappeared as testimony to the events of 1984. They testify to—or haunt—the official denials and narrative of Bhopal. The Bhopalis are infused with ghosts and the memory of the disappeared. This enables the forging of a local cultural identity (gas victims) *through* the figure of the missing or the dead, and makes Bhopal the spectral city. Haunting generates identification and communitarian definition in Bhopal.

The spectral city is a palimpsest where the memories of death, disappearance and helplessness are layered over with projects to recover, rehabilitate and repair. It is here that the next instantiation of the haunting of Bhopal's urban nature may be found.

Prajapati in his account describes the government initiatives to develop "'green belts," additional low-cost housing, industrial units ("work sheds," as he calls them 114), training institutes, community centres, "asphaltation of roads" and other urban development initiatives (Prajapati 110–130). Such state efforts to alter the landscape of disappearance and erase the manifest absence through construction and urban planning are at odds with the issue networks" emphasis on the aid required by the victims. As early as 1985, a year after the disaster, the Rashtriya Abhiyan Samiti fighting for justice for the victims, criticized the government efforts to erase the landscape of disappearance and "beautify" Bhopal. In its 1 September 1985 press release the Samiti wrote:

> The Samiti has stated that the much-publicized Rs 322 crores scheme in fact makes a provision of only Rs 38 crores for the rehabilitation of gas victims. The scheme is actually an unabashed attempt for the so-called beautification of Bhopal. . . . Only the Government and its experts can explain how an aerodrome, a railway station, a botanical garden, a boating facility and a swimming pool etc. are related to the rehabilitation of gas victims. . . . (APPEN 85)

Lapierre and Moro note that in 1997 the Indian government's "minister for culture suggested turning the whole of the Kali grounds into an amusement park" (361).

The Rashtriya Abhiyan Samiti press release draws attention to the attempted erasure of memories, of transporting landscapes of disappearance into landscapes of recreational and tourist spots. Such an alteration of the landscape of helplessness and disappearance also generates the Bhopal uncanny. The non-absent past has to be rendered totally and irrevocably absent from the present by erasing all signs of this tangible past.

Robert Mighall writing about urban Gothic argues that "for the urban Gothic this meant the criminal past haunting the civic present" (2007: 55). He adds: "The premise of Gothic fiction, dividing the civilised from the barbarous, the progressive from the retrograde or anachronistic, is here located in the metropolis of the modern world" (55). This "incongruity" (Mighall's term, 55) of the two worlds within the same "space" that is Bhopal generates its uncanny. To return to Mighall's elaboration of the urban Gothic/uncanny:

> So what makes the urban Gothic? For Gothic *of* a city rather than just in a city, that city needs a concentration of memories and historical associations. Ideally these would be expressed in an extant architectural or topographical heritage, as these areas provide the natural home for ghostly presences of imagined/projected meanings. (57, emphasis in original)

The architectural and topographical changes proposed, initiated and executed by the state build over the public memories that constitute the city. When Francesca Moore photographs miles of the UCIL factory wall (*Bhopal: Facing 30*) she is attempting to *retain* the space of loss and disappearance, as well as the melancholic public sphere around these spaces. Prajapati's photographs of new structures, roads and green belts are the incongruities when juxtaposed with Moore's wall. Moore captures the "concentration of memories" that Prajapati's images ignore or write over. If, as argued above, Bhopal and Bhopalis build an identity around the crypt enclosing absence and loss, then these projects of urban revival are attempts to erase historical associations and memories. The uncanny emerges from this manifest absence that seeps through, in the form of, say, Moore's images sitting incongruously adjacent to the new landscapes.[2]

Necropolitics of the kind Perera proposes via Mbembe also consists of erasing memories, narratives and spaces of the dead. That is, the biopolitical uncanny in the cultural texts gestures at the state-sponsored yet corporatized necropolitics which, first, denies the number of the dead and the disappeared and second, effaces and erases spaces that are *about* the dead and disappeared.

Beyond the uncanny emerging out of the necropolitics of erasure of spaces of mourning and memory, of the material signs (like photographs) of the missing and the dead, there exists in the Bhopal cultural texts another form of the uncanny, of the persistence of the past.

"SLOW VIOLENCE"

Bhopal is often described as a "continuing disaster." This trope of an extended, unending and ongoing disaster works to describe the ongoing pollution of the natural resources and environment *and* to describe the corporeal and medical trauma of survivors. MIC and assorted toxins are the hyperobjects that have adhered to soil particles, water bodies, human tissue, from the visible to the invisible (molecular levels). MIC is no longer "available," but like the extended half-life periods and non-decomposing products of human industry such as plastic, MIC simply refuses to go away. As a hyperobject it exists beyond time and space, is nonlocal and adhesive in its occurrence, iteration and effects. MIC as a hyperobject is the ultimate foreign body that haunts.

The poisoning of bodies, plants, land and water with deadly elements like cyanide in the form of delayed toxicity is captured in the Indian Council for Medical Research's bland statement in its 2010 report: "this study had brought to light the mechanisms of acute, delayed and recurrent cyanide toxicity" (Bhopal Gas Disaster Research Centre and Indian Council for Medical Research 2010: 142). So: acute, delayed and recurrent—this is the slow violence of Bhopal.

This poisoning, incidentally, is *not* the effect of the events of 1984, but has a prehistory. In his letter (already cited in chapter 1) to UCIL of 4 March 1983, advocate Shahnawaz Khan wrote:

> That waste water that contains poisons flows through a pipe line going through the middle of Jaiprakash Nagar Colony and falls into the sewage canal near the crematorium. This poisonous water poses danger for the residents and cattle. Poisonous water going into the earth from this pipeline and traveling through aquifers is contaminating the wells in Jaiprakash Nagar and hand pumps and tube wells in the area. Because of this the well water in this area cannot be used for drinking. (Hanna 2015: 19)

Gothic texts are marked by violence, of various kinds and of varying degrees. In the case of Bhopal's ecological Gothic this violence takes on specific forms. The true horror of Bhopal's Gothic is *not* only the events of 2/3 December 1984 but the continuity and persistence of the events in terms of their effects. In this Bhopal's Gothic approximates to both Rob Nixon's (2011) account of environmental disaster as "slow violence" and Adriana Cavarero's argument about "horrorism."

Nixon proposes that "slow violence" is "violence that occurs gradually and out of sight, a violence of delayed destruction that is dispersed across time and space, an attritional violence that is not seen as violence at all" (2). It is "neither spectacular nor instantaneous, but rather incremental and accretive, its calamitous repercussions playing out across a range of temporal scales" (2011: 2). Such slow violence affects the environment and the human body in incremental fashion, is transgenerational and across time frames. Slow violence effects a horrorism. Adriana Cavarero defines the effect of horrorism thus: "a body reduced to a totally available object, or, rather, a thing objectified by the reality of pain, on which violence is taking its time about doing its work" (31). It is a violence that "not content merely to kill because killing would be too little, aims to destroy the uniqueness of the body, tearing at its constitutive vulnerability." It is about the prolonging of suffering, in Cavarero's reading, as instanced in the case of Mitya Bai, a survivor whose account appears in the *777* newsletter: "From the moment she breathed the gas Union Carbide said was 'harmless', Maya never knew another day free of pain. The poisons took 16 terrifying years to kill her." (*777*, Sept. 2004)

Toxic Haunts

> But [on] the contamination issue—things are even sadder—because the government does not recognize them [people exposed to environmental con-

tamination] as people affected by Union Carbide's poisons. (*Amnesty* 2014: 12)

> The results of this survey indicate general contamination of the site and immediate surroundings with chemicals arising either from routine processes during the operation of the plant, spillages and accidents, or continued and ongoing release of chemicals from materials which remain dumped or stored on site . . . the survey conducted by Greenpeace International has demonstrated substantial and, in some locations, severe contamination of land and drinking water supplies with heavy metals and persistent organic contaminants both within and surrounding the former UCIL pesticide formulation plant. (Greenpeace, *The Bhopal Legacy* 1999: 3–4)

The first quotation demonstrates, in one crucial turn of phrase ("does not recognize them as people affected by Union Carbide's poisons"), what Rob Nixon has argued: that the poor are the ones most affected by slow violence: where "their unseen poverty is compounded by the invisibility of slow violence" (4). The second quote describes the state of the soil, water and other resources decades after 1984. The hyperobject that is MIC extends beyond the date/year chronology.

What haunts Bhopal and Bhopalis is the landscape of fear, polluted, contaminated and killing them slowly. In her reading of Pat Barker's fiction, Amna Haider argues:

> Fear in this context is inextricably linked to control and possession over space . . . [generating] questions of possession and dispossession, empowerment and disempowerment over another crucial site, in particular the psychological space. It is important to note how landscapes of war overlap and colonize the landscapes of the mind . . . abodes that are transformed into the loci of otherness for the self in the Gothicized inverted world. . . . (2012: 56–57)

This is a Freudian uncanny for, "the 'Uncanny' lurks in the space opened up by trauma wounds, where the known coalesces with the unknown; the familiar with the unfamiliar and the safe with the unsafe" (57). Haider points to the transformation of the known and the familiar into strange, frightening spaces that then produce psychological trauma, spaces in which one feels oneself *as* the other. Just as the victim or survivor's body begins to be experienced as foreign due to the poison inside it, the landscape of Qazi Colony and other areas adjoining UCIL's lands are experienced as foreign due to something in the air, and in the water. Homesteads are stranger shores after 1984.

Take, for instance, the stories of dangerously contaminated water in the region around the UCIL plant over 30 years after the disaster:

> Atal Ayub Nagar—a dry, dusty strip of land sandwiched between the main railway line and the walls of the abandoned factory. The people here are poor

> and the houses are mostly built from wood, plastic sacks, cow dung, corrugated concrete and metal sheets. On my last visit I was told that, as the water tanks were only filled sporadically, many of the families were forced to use one of the two government hand pumps. They complained that the water smelled of chemicals and had a bad metallic taste but they told me they had no choice but to drink it. . . . Our own B[hopal] M[edical] A[ppeal] survey showed that levels of carbon tetrachloride alone were 4,000 times higher than the World Health Organisation's safe limits. (Toogood 2011: 8)

Or, as a character in *Animal's People* puts it: "our wells are full of poison. It's in the soil, water, in our blood, it's in our milk. Everything here is poisoned" (107–8). That is, the nature of their setting, of their homes, is what kills the survivors, slowly. UCIL haunts them in the form of embedded poisons.

In other cases, we see a displacement and dispossession from their homes, employment and social relations as a result of the events of 1984:

> Thousands have lost their livelihoods. Many are now homeless beggars. Their breathless bodies no longer able to push handcarts and lift heavy loads, they have fallen into utter destitution. (*777*, 2003)

The survivors are made stranger to themselves due to the kinds of physiologies and anatomies they are born with. Reports of monstrous and deformed births since 1984 abound in various studies, all indicating the lingering trauma of those haunted by 1984. That is, those who are residents in the UCIL area are rendered foreign to the place because it is no longer "home," and foreign to their own bodies because most of them no longer fit the norms of being "human." This is the estrangement of humans brought about by the slow violence of Bhopal's soil and water. While MIC is the interobject occurring between chemicals, objects and humans, it relegates humans to nonfunctioning or quasi-functioning bodies.

Their homes are places of breakdowns, diseases and loss. Their community is the space of destruction, dispossession and displacement from employment, safe environments and a sense of security. Slow violence here is the complete helplessness and lack of control over their home and community spaces. The relations of the residents to the landscape they occupy is one of fear and distrust, years after UCIL has vacated the place.

What I wish to draw attention to in these texts is the particular *kind* of Gothic monstrous. There is no external referent or monster, no alien or strange visitor. Rather, the monstrous is within the home, the soil, the water and the body. It is internal to the operations of domesticity and community. It is the everyday setting that is Gothic, haunted by something that cannot be seen but can only be experienced as fear when it manifests in the form of diseased and deformed bodies and ongoing pain. The twisted body of "Animal," the chief protagonist of *Animal's People*, is a symbol of all the humanity that burnt out that night, and the misshapen state of those who survived.

The alien has been internalized, even as the home and the community has been rendered an alien landscape generating fear rather than security. Toxic haunts via slow violence is the slow inversion of spaces of safety and security, and generates the uncanny of strange homes and stranger communities occupied by people whose bodies have become strange and alien to themselves.

Slow violence as Nixon argues is often invisible and its victims, being poor, are invisible too. The source of the residents' diseases, mutations and fears are invisible, microscopic and chemical in *nature* (I use the term in its double sense of "essential quality" and natural setting/environs). In one sense UCIL has not left the region: its legacy is the haunting, the slow violence.

If the uncanny is "not a property of location alone but highlights the power of certain locations, of sites of trauma . . . to imprint themselves on familiar terrains" (Haider 63), the slow violence in the residential zones of Bhopal imprints the landscape upon their minds as alien. Homes and communities are places that remind the residents constantly, as the interviews and reports show, of their losses. Toxic haunting, to phrase it differently, is the poisoning of the survivors" sense of place.

This poisoning has one more dimension to it: the slippage between the natural and the man-made proceeding from the slow violence. First, natural resources such as ground water and soil are contaminated beyond repair, as the reports uniformly show. This contamination is man-made. Second, resources supplied by/from the state in the form of piped water, for instance, are inadequate and drive the residents to using the contaminated water. Thus, the slow violence is not merely the poisoning of natural resources from UCIL but the continual neglect of ensuring safe, and adequate quantities of, drinking water to the neighborhood. If the first clearly demonstrates the man-made destruction, through invisible contaminants, of natural resources, the second demonstrates the violence of the state and state policies that condemns the survivors to continuing trauma. If nature has been polluted by man's efforts, then other such human efforts drive the survivors to that same nature.

The debates around the exact composition of the gas, the contaminants and the extent of pollution in both material bodies and the environment reinforce the hyperobject nature of MIC: it is everywhere as its material effects.

The Molecular Family and Haunting

I have already examined the family portrait of Bhopali survivors above to argue that such portraits are built around visual ellipses—suffering, which does not find a place in the photograph—but which defines the domestic scene because the suffering is part of their domesticity and identity. This identity is one they are aware of as embedded deep in their genes: children born with birth defects because their parental bodies are reservoirs of mutated genes and poisons.

Moore's portraits (in *Bhopal: Facing 30*, 2013) might capture the nuclear family in most cases or even the extended family in others. Interviews with survivors tell us of their *family* medical conditions: each family aware of their genetic, that is inherited and transmissible character. This is the "molecular family," wherein genetic relations are now deemed to be foundational to familial relations (Couser 2004: 177). Bhopal's molecular families are the ones aware that their family relations are defined by the poisons inside them, by their damaged genetic code, even if they do not understand genetics.

"A woman, Jameela, confides that 'too many children' are being born with birth defects," says a report (*777*, April 2004). Another gestures at the slow violence of molecular families:

> In a tiny house, spotlessly clean, you meet the parents of Iqbal and Zarina, both of whom are in their teens and have congenital malformations. Says their mother, "Union Carbide will live forever, but we will not." (*777*, April 2004)

Or another:

> Children exposed to the gas *in utero*, or born to gas victims, have suffered from stunted growth, deformed limbs and mental retardation, cleft palates and blindness. Growth retardation in adolescent boys is also widely reported. (*777*, 2014: 4)

Slow violence is the process by which these individuals and their families cast their family histories and life stories as biological data and genetically transmissible conditions. Individuals document risk-as-life and life-as-perpetual-risk in this process of examining their lives.

Further, when we examine the life narratives of the survivors, they explain how they have fashioned their lives around their medical conditions. I have elsewhere argued that in the "genetic citizenship" (Heath et al. 2014) and heightened knowledge of their genetic conditions and predilections to certain illnesses, individuals offer life narratives around this medical knowledge.

> What emerges in the life writing of such a self-aware individual is the predictive self. The predictive self is fashioned out of the genomic data that are now available. Individuals speak of themselves as being in a state of embedded vulnerability, selves that can only perform the genome, so to speak. Embedded vulnerability has to do with the state of being powerless in terms of life choices, given the prediction and determination of features and futures by genes. (Nayar 2016: 9)

Survivors speak of the kind of employment they can undertake, their visits to hospitals and their search for palliative relief, cure or simply death. Here are two instances from Amnesty's *Clouds of Injustice*:

> Speaking about the impact of poor health on herself and her five children, 46-year-old Hasina Bi from Atal Ayub Nagar said: "Since birth I have worked—I used to roll *beedis* [cigarettes]. I have problems with my vision so I stopped. . . . My children can only work for two days at a time [due to poor health]. . . ."
>
> Mohammad Ayub, 45, from Jai Prakash Nagar, has been unable to work as a driver for long periods due to severe pain in his hands and shoulders. His family has incurred large debts as a result. (Amnesty 2004: 19–20)

Another study published in the respected *Journal of the American Medical Association* in 2003 states unambiguously: "We found selective growth retardation in boys, but not in girls, who were either exposed as toddlers to gases from the Bhopal pesticide plant or born to exposed parents" (Hanna et al. 146).

These are haunted families, with life choices such as employment, education, marriage, child-bearing determined by the medical condition generated by the poisons either inhaled (on 2/3 Dec. 1984) or imbibed through the inheritance of damaged genes. Molecular family narratives such as the above are about this embedded vulnerability that determines their life choices, their very sense of self and the sense of being members of a family. Molecular families waste away in this fashion, unable to work, marry, lead a normative family life, fulfill their aspirations. Haunted by their dead relatives, friends and neighbors, they continue to imbibe more industrial waste through the slow violence of their environs.

Waste threatens "the very integrity of our body politic" (John Scanlan cited in Morrison 2015: 63). It is therefore "a form of the uncanny" (Morrison 63). Just as UCIL's wastes and remainders of the plant haunt Bhopal and its survivors, the survivors are sadly wasted bodies that haunt the city and its neighborhood. MIC is the agent that is produced as a result of a chemical-industrial process but which exists, modified, metamorphosed, in some form at the molecular level inside the human. The deformed, sick and suffering bodies of survivors and the children of the exposed might then be treated as embodying the invisible yet material hyperobject that is the MIC. MIC extruded and exploded as waste matter that night in 1984, but the material persistence—the sheer viscosity of MIC that binds it to the molecules and tissues of humans *across* generations and therefore time—of the chemical inside the bodies ensures their wasting away. As waste matter, MIC possesses *agency*—insofar as it disturbs the agency of the human body, and renders it waste/d. In its interobjective quality, it attaches itself to the human tissue and chemicals, and asserts its agency.

WASTED/ING HUMANS AND THE HAUNTED BODY POLITIC

The repeated campaigns around the wasted and wasting bodies ensure that the theme of waste haunts Bhopal. It has become impossible to speak of Bhopal without a reference to the continued wasting away of its people and citizens. That is, "gas victim" becomes the nomenclature for not only citizenship but also for identifying Bhopal as a body politic. If waste accumulates and contaminates the body politic as Susan Signe Morrison's work (2015) demonstrates, wasted/ing bodies—that are themselves the haunts of poisons—threaten all image-making exercises of Bhopal.[3]

Wasted/ing bodies have also haunted the body politic of transnational events like the London Olympics, as we have seen. Protests against Dow went international and aligned local waste with global concerns of slow violence. The Anti-Hu campaign of Bhopal.net emphasizes this global haunting of the body politic via wasted/ing bodies.

In this poster against Dow's own advertisement campaign "The Human Element," Bhopal's slow violence is linked to the company's "chemistry"

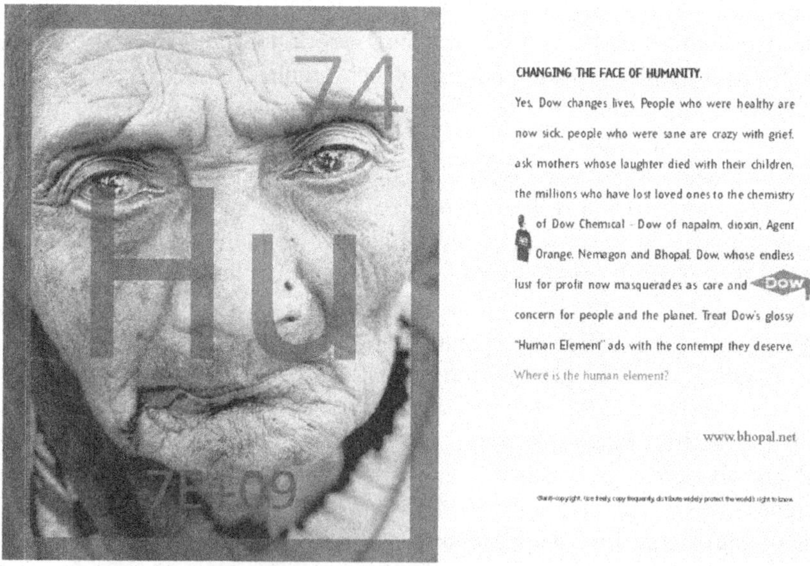

Anti-Dow campaign poster. Courtesy International Campaign for Justice for Bhopal.

globally: napalm, Agent Orange and dioxin. It links various places and disasters: Nemagaon, Vietnam, Bhopal. It suggests that Dow is a universal enemy to all humanity. It is a transnational haunting by the poisons exuded and disseminated by its plants all over the world. By linking the various products and places the above campaign suggests that Dow's poisons are not only about individual bodies being "wasted," but rather about the poisoning of entire populations and the body politic.

The biopolitical haunting of Bhopal works on two key fronts: the community haunted and poisoned, now and for all time, by the invisible *material* detritus—in the form of poisons in the air and water—of UCIL's processes and the discursive poisoning in the form of accusations that the community itself produced the workers that then wreaked havoc through sabotage. If the former focuses on wasted/ing *humans* whose presence troubles the composition of the Bhopal society today and tomorrow by pointing to ongoing abjection and trauma, the second gestures at the *social pathology* inflicted upon the fabric of the community in the form of accusations, suspicions and criminal charges.

Waste, pollutants and contaminating poisons haunt the bodies of individuals, but also the body politic of Bhopal and the various cities in which UCIL/Dow has manufacturing units. Unlike in traditional Gothic texts where the family and the home are the loci of decay, in the ecological Gothic of Bhopal's cultural texts it is the city itself that is decaying and the community as a whole that is being poisoned by UCIL's debris deep inside its water and soil. The dependence of the communities and neighborhoods upon polluted water, the scenes of children playing in abandoned yet poisoned UCIL lands are two regular images in the cultural texts and capture this theme of haunted and decaying body politic, or a social pathology. Bhopal's biopolitical uncanny moves, then, from wasted/ing individual bodies to families and the community at large, thereby also suggesting the collapse of the body politic due to the entropic diffusion of UCIL's persistent poisons in the environs. Here the Gothic is no longer exogamous—for, as argued in earlier chapters, UCIL is the foreign inside the self—but intrinsic to the community and its landscape. This foreign alters family structures and community demographics by altering reproductive rates, the number of normative and productive human bodies. For instance, Amnesty's 2014 report, *30 Years Is Too Long*, cites:

> The experience of Shahzadi Bi's daughter, who was born after the disaster, is one of many such cases among the next generation cited by the community. "My daughter couldn't conceive for four years after her marriage," she says. "Doctors had told her clearly that 'since you have been drinking this toxic water, you will not be able to give birth.'" (Amnesty 2014: 6)

In Rahul Varma's *Bhopal*, the woman Izzat, the mother of a severely sick infant, Zarina, is asked by Devraj, the Head of Carbide International in Bhopal: "You give me the names of the mothers who visit this Doctor Sonya for their sick children." Izzat responds: "Here. Veena, Budhiya, Kasturi, Shanta, Farida, Phoolmati, Babban, Meeta, Imarti, Roshni, Rani. . . ." (20014: 8). The list, as the ellipsis suggests, is not complete, and gestures at the numerous, ongoing haunting of wombs and bodies. Sonya informs Madiha, who refuses to believe that Carbide is poisoning wombs, including her own, later in the play: "Bhopal is being littered with deformed and stillborn babies" (53). Zarina in Varma's *Bhopal* is described by her mother Izzat at the child's grave: "Zarina died the day she was born. One leg, one arm, no fingers . . . there was nothing right with her. Her tiny little heart was like dried rubber. . . . I could see it under her skin" (64).[4]

Raghu Rai's photograph of bottled fetuses (collected by Dr. Satpathy)—aborted by pregnant women caught in the gas leak—embodied lives never lived (Rai, *Exposure*, www.magnumphotos.com). Amar, Anjali and Sandeep's son in Amulya Malladis's *A Breath of Fresh Air*, is born with congenital heart and lung disease. Malladi writes: "Anjali hadn't known that the effects of that deadly night in Bhopal would lead to a child with a weak heart and weak lungs" (2003: 62). Amar dies at the end of the novel, at the age of 12.

This emphasis on deformed, stillborn, dying babies in Varma's play, Malladi's novel and the other texts resonates with a Gothic theme in horror writing. Dawn Keetley, who notes that such an "animus" toward children is a "persistent thematic of Gothic fiction" (2013: 93), in her analysis of the TV series *American Horror Story* argues that "the deaths of babies and children preoccupy *American Horror Story* and are a powerful sign of its vision of a failing futurity" (93). Bhopal is denied its futurity, thus once again telescoping past with the future in the form of the haunting. The haunting image of deformed and aborted fetuses (in Raghu Rai's photograph) is echoed through the campaign writings, the visuals and the poignant Indra Sinha novel, *Animal's People*. If the Gothic is characterized by a "fearful sense of inheritance" (Baldick 1992: xix) then the inheritance of the Bhopalis is exclusively in the form of mutated babies and still-borns.

The fetus in the jar talks to the boy-narrator ("Animal") in *Animal's People* (57–59, 236). One of the fetuses claims that they are the Board of Directors of the Kampani:

> Everyone on this earth has in their body a share of the Kampani's poisons. But of all the Kampani's victims, we are the youngest. We unborn paid the highest price. Never mind dying, we never even got a fucking shot at light. This is why, Animal miyan, we are the Board of Directors of the poison wallah shares. (237)

The ones who never lived haunt the premises in the form of the unborn (as opposed to the undead). They are the primary victims, or even the ultimate victims, of the events of 2/3 December 1984. In terms of the "shares"—a term that in its cruel irony encompasses both the financial and the "lot" that fell to the fetuses—from UCIL, they did receive the highest and most pernicious share of the gas that leaked. That is, the unborn corner the market, literally and corporeally, in terms of the output from the factory. The fetus is victim, witness, shareholder, all rolled into one, in the toxic economy of Bhopal.

The spectacle of the deformed body and the aborted fetus in Raghu Rai and Indra Sinha offers us another route into the haunting of nature in Bhopal's cultural texts. Alexandra Schultheis Moore reads the representation of Animal in Sinha as "resignifying subaltern and disabled people" by making "visible the interplay of structures of (neo)colonial power, juridical-medical discourse" (2013: 238). Rosemarie Garland Thomson famously argued that:

> Constructed as the embodiment of corporeal insufficiency and deviance, the physically disabled body becomes a repository for social anxieties about such troubling concerns as vulnerability, control, and identity. In other words, I want to move disability from the realm of medicine into that of political minorities. . . . (1997: 6)

Thomson, whose work Alexandra Moore (cited above) also employs, enables us to see Animal not as a medical anomaly but as a corporeal instantiation of a series of processes: the poisoning of bodies by MIC, the absence of accurate medical knowledge, the lack of support structure and finally, poverty. When Animal declares "I used to be human once" (the first line of the book), he refers to not just his physical disfiguration but the very identity and ontological status that has been disturbed by the events of 2/3 December 1984. What haunts Bhopal's nature, including land, earth, soil, bodies, is the reconfigured body that can no longer be associated with the term "human." Animal is not a deformed body: he is a spectacle that recalls an entire narrative, from misinformation to poverty, as it is inscribed upon his body.

Raghu Rai's photograph of the Satpathy exhibition of spontaneously aborted fetuses from that night marks a convergence of the medical and the media. On the one hand they are a medical spectacle, but on the other they perform a key cultural role: of information, entertainment and commerce (I adapt here the work of José Van Dijck on the medical documentary, 2002). It takes a specific medical condition and transforms it into a public spectacle. The fetuses are medical anomalies but they are also "things" that mark the boundary between the normal and the pathological, a boundary that was put in place by the events of 2/3 December 1984. Rosemarie Thomson argues that "extraordinary bodies [were not only] a source of profit, but the narratives of pathology derived from monstrous bodies built reputations" (58). In the case of Satpathy's embalmed foetuses the narrative of pathology highlights the reputation of UCIL. They serve as reminders (and remainders) of technological (machinery failure), cultural (misinformation) and medical (poison gas) discourses *and* events that came together to "produce" these anomalies and pathologies. In other words, the fetuses are not medical anomalies alone, they are the product of a chain of processes that involve technology, culture and medicine. Staging them as spectacles, as Dr. D. K. Satpathy and Indra Sinha do enables the medical anomaly to be located within these multiple discourses. The "deviant" body in these cases is not simply a medical aberration. This is an embodied "true life" story that is at once advertisement and souvenir (Thomson 61) but also a memorial. The fetus is no longer a freak of nature but a freak of culture, as Thomson (via Susan Stewart) reminds us (62).

The display of fetuses combines medical curiosity and anomaly with the cultural anxieties of lives cut short. The fetuses emblematize the fate of Bhopalis: a fate worse than medical death, because death implies a life lived before termination. In medical history, "exhibited human remains were above all those of criminals and these emblems of the material culture of medicine were often used as moralizing vignettes" (Talairach-Vielmas 2015: 33–34). These medical displays were incorporated into the Victorian Gothic literary-cultural texts, often to sensationalize medicine, crime and the body, as Talairach-Vielmas persuasively argues. In the Bhopal story, the fetuses have a similar effect: they generate the uncanny which is part of the ecological Gothic.

The fetus embodies a future that will never be. Blameless victims, such as the fetus, serve as potent symbols of toxicity carried through bodies into molecules, genes and therefore *beyond* individual and generational boundaries. As an assertion of the biopolitical uncanny, reproductive failure embodied in the fetuses, haunts the modern Indian world itself: the fetuses remind us of the failure of the project of modernity which was UCIL. That is, the damaged and dead fetuses symbolize the crisis of modernity which works at the level of both production (industrial) and reproduction (human). Both production and reproduction are connected in that the excesses in one (industrial production) leads to a diminishing in the other.[5]

If the fetus embodies the crisis of India's industrial modernity and the biopolitical uncanny, a different form of this uncanny, connected to the thematic of production, may be discerned in the industrial *worker* of UCIL.

The Indian workers of UCIL were blamed for the events of 2/3 December 1984 (Saxon 1986, Weisman and Hazarika 1987). The Ashok Kalelkar–Arthur Little report stated that

> a disgruntled operator entered the storage area and hooked up one of the readily available rubber water hoses to Tank 610, with the intention of contaminating and spoiling the tank's contents. It was well known among the plant's operators that water and MIC should not be mixed. He unscrewed the local pressure indicator, which can be easily accomplished by hand, and connected the hose to the tank. The entire operation could be completed within five minutes. Minor incidents of process sabotage by employees had occurred previously at the Bhopal plant, and, indeed, occur from time to time in industrial plants all over the world . . . those involved decided upon a cover-up. They altered logs that morning and thereafter to disguise their involvement. (1988: 15–16)

Jackson Browning's report complained that the UCIL (USA) team "was hampered in its work by the Indian Central Bureau of Investigation, which had taken control of the plant" and "the Indian government prevented access to witnesses and records in India" (1993: 11). It then concluded: "evidence showed that an employee at the Bhopal plant had deliberately introduced water into a methyl isocyanate storage tank. The result was the cloud of poisonous gas." (11)

In Varma's *Bhopal* Anderson smoothly shifts the blame to the Indian workers and management: "What do you people get up to in this country?" (48). Then adds in fury: "I and my men and women have hung on to a single hope—that Carbide would be a good citizen in each of those nine hundred towns [which have Carbide plants], and, as in Bhopal, would help and put an end to poverty. Help realised untapped potential. And you, Mr. Sarthi, have killed that hope" (51). Devraj Sarthi clarifies that the key refrigeration plant was turned off at the request of the Carbide headquarters (48)—a point Anderson simply ignores during his rant.

The sabotage theory therefore tracked the origins of the horror and the poisons to people who technically "belonged" to the plant. Indeed, Chouhan's *Bhopal: The Inside Story* (1994) was published precisely with the aim of countering these charges from UCIL. The monsters, in this accusatory rhetoric from UCIL, come from within Bhopal's own community and UCIL workers. Such an accusation points to a decaying social fabric within the UCIL *India* Limited (and by extension Bhopali) community: disgruntled workers and disaffection as a cause of the horror. What UCIL's rhetoric achieved was to paint a corrupt Indian working class community. The evil, it suggests, was of India's own making.

The biopolitical uncanny as constituted within these two moments exhibits a specific social or cultural uncanny. The manner in which "Carbide" or "MIC" circulates in the cultural texts suggests a revenant who determines and defines the identity of Bhopal and Bhopalis. Carbide and MIC are specters haunting, and hence "housed," in the lives and minds of the characters. But this alone does not construct a cultural uncanny. The uncanny was linked, Allan Lloyd Smith argues, to "specific features of culture, such as imperialism and the fear of what is brought back from the colonies" (Smith 1992: 285). In Bhopal's cultural texts the cultural uncanny is the scattered remnants of such "foreign" features from cultural and national pasts within the bodies of survivors. If the uncanny is, as Freud defines it, "that class of the terrifying which leads back to something long known to us, once very familiar" (369–70), then UCIL, MIC and the plant itself terrifies by being a continuing presence in the rhetoric, settings, bodies and lives of the people. The cultural uncanny calls for a repetition of the past—in this case, a repetition that is inseparable from the cycle of violence UCIL unleashed then and continues to unleash now. UCIL and MIC become the *foreign* within the people. It is a shared foreignness because it returns spectrally from a community's collective past. If the past is a famous "foreign country," then UCIL is the foreign element who/which is inside the consciousness of Bhopalis and Bhopal.

These two dimensions of a cultural and communitarian haunting therefore have two apocalyptic forces in conjunction: the elemental, material poisons and the discursive poisoning of the tragedy by UCIL's lawyers and representatives. But it is possible to discern a third apocalyptic force in the continued haunting of Bhopal: the sense of fear, suspicion, vulnerability and helplessness that demoralizes the community as a whole.

Brian McDonald working on Don deLillo's post–Cold War Gothic argues: "paranoia, suspiciousness, and a general feeling of insecurity are . . . a significant part of the residual psychic waste of the Cold War, and indeed waste is the other implacable Gothic force that haunts. . . ." (2008: 99). This psychic waste is, in the case of Bhopal, conjoined with the material wastes, of course, as I have pointed out above. But the fears set in motion renders the psyche, individual and communitarian, wasted as well.

We detect this third apocalyptic force, of psychic wastes/wasting, at several moments in cultural texts. The first is clearly in the immediate aftermath of the disaster where secrecy, ignorance and the ill-equipped state drive the injured citizens into fear and helplessness. Anees Chishti records children wailing for parents, the injured wailing for medical relief—indeed, any relief—from the pains and the sheer lack of adequate equipment to deal with the magnitude of the catastrophe (28–29). Thapa describes "many people running crazily, looking for their family members in tears" (chapter 2, unpaginated). But we also see this apocalyptic, crazed sense of fear and helplessness in the aftermath of the compensation ruling by the legal system. As Safreen Khan puts it in the Amnesty Report *30 Years Is Too Long*:

> People have run out of patience. They still remember. They still cry and mourn for their family members who died that day. They feel that at least now, our

> government and the company must listen and take steps, because 30 years is too long . . . to get justice. . . . (Amnesty 2014: 2)

We see it in the despair of the survivors today when medical aid, information (about what exactly is poisoning them), basic amenities are all lacking. In other words, I see the biopolitical haunting as also the effect of this third apocalyptic force that informs the people and community. Such a force gives shape to a popular and communitarian consciousness that becomes obsessed with secrecy and misinformation as the root of UCIL's continuing power in/over Bhopal, and with helpless rage as the only response of the already vulnerable. The last word on this thematic of speaking, information and promises must necessarily belong to "Animal": "On that night it was poison, now it's words that are choking us" (Sinha 3).

NOTES

1. Hirsch writes: "Like all pictures, the photos in *Maus* represent what no longer exists. But they also represent what has been, and what has been so violently destroyed. And they represent the life that was no longer to be and that, against all odds, nevertheless continues to be. If anything throws this contradictory and ultimately unassimilable dimension of photography—its hovering between life and death—into full relief, it has to be the possibility, the reality, of survival in the face of the total death that is the Holocaust" (9).

2. Following Anthony Vidler's work (1992) on the "architectural uncanny" it is also possible to see how the new Bhopal landscape and structures defamiliarize the familiar landmarks, notably UCIL and its environs.

3. Sara Wasson argues that in contemporary Gothic science and literary fiction such as Ishiguro's *Never Let Me Go* we can discern a thematic of bodily organs as waste. Wasson's larger point is that "dominant cultural metaphors of organ transfer may shape institutional processes of tissue management, and how medical protocols of tissue procurement may become disturbingly influenced by capital's imperatives" (2015: 105). Wasson's alignment of the state's role with capitalism's market economy, and the effect on organ harvest and trade might also be read in terms of the collapse of cultural protection which directly impacts the corporeal.

4. NR Bhandari, the Medical Superintendent at Hamidia during the disaster, provides data on the surveys conducted of women in the gas-affected areas. He documents: "By 1st September 1985 every pregnant women [*sic*] who was exposed to gas had delivered. . . . Total of 2698 women were pregnant at the time of gas leak. Out of these 2210 gave birth to live deliveries, 86 were still born and 402 cases of abortion were noted. . . . There were 19 cases of twin deliveries. 30 infants had congenital malformation of which 7 cases had minor malformation and 23 cases had other malformations. . . . 378 women had spontaneous abortions . . . out of the 2210 live born infants, 150 died, 78 infants died in the first 7 days of their life where as 36 infants died at the age of 8 days to 1 month. . . . The prenatal mortality in the MIC affected population was 71.4 per thousand live births" (Bhandari 2010: 65–66). Another study, by Daya Varma in 1987, recorded: "43% of pregnancies in women residing near the Union Carbide pesticide plant did not result in the birth of a live child" (Hanna et al. 127). Evidently, there is a crisis in reproduction that all accounts of Bhopal document.

5. Critics have noted that modernism as a literary movement in Europe itself might be read through the lens or a "crisis in reproduction" (see, notably, Hauck 2003).

Chapter Five

Bhopal's Precarity

Toxic History and Thanatopolitics in the Postcolony

According to the data with the Government of India, there are 1252 square kilometers of industrial ruins across the country (with data "Not Available" for four states from Northeast India, Tamil Nadu and the Union Territories), of which perhaps, the most famous will remain, for a long time, Union Carbide's ruins in Bhopal.[1]

This chapter reflects on the post-disaster, survivor Gothic of Bhopalis. It examines the role of derelict but haunting, empty yet symbolically loaded spaces of the UCIL plant, its environs but also the Bhopal landscape peopled by the injured bodies of survivors, by annual assemblies of these bodies as, thirty years after the disaster, the campaign to recognize the injuries and injustice done to them continues.

The Gothic's "remains" are material—predominantly landscapes of ruin and waste, of both buildings and humans. This itself is a sublime of wastes that signifies a collapse of civilization and modernity. Unlike the traditional Gothic with its insistence on the supernatural and demonic that produce wasted beings and wasted lands, the Bhopal Gothic, like the dystopian landscapes in numerous contemporary cultural texts, is man-made. The postindustrial ruin instantiates a "decadent sublime" (Presto 2011). Postindustrial landscapes serve as the raw material for art in the form of dystopian films in the late 20th and early 21st century in films such *World War Z, The Book of Eli, 2012, I Am Legend, Mad Max: Fury Road, 28 Days Later*, or photographs (Edward Burtyns, for example). Waste, writes Patricia Yaeger, is "an alternate site of reading history," a space where the teleology of modernity has finally arrived (2003: 106). Yaeger's argument enables us to see how devastated landscape are also material signs of a certain history. If haunting, as we have argued, suggests the persistence of poisons, memories and slow violence in the Bhopalis, the devastated land signifies the climax of a particular history—of industrial modernity and its unregulated production of Sevin.

BHOPAL'S POLYCHRONICITY

In the Western scenario, these derelict spaces are transformed into heritage sites, as Tim Edensor points out, and thus conflates past and present into a consumer product (see chapter 5). This is emphatically not the case in India.

Strongly reminiscent of the Gothic's formulaic representations of ruined abbeys, castles and houses which induced terror and awe, and which was the space of the ancient ghost or evil meeting the present, the industrial ruin mixes up temporalities. Disused machinery gestures at the power the space used and generated and products consumed and produced. That the larger frames of the machines remain tethered to the ground gestures at the machine as temporal artefact fusing past and present, or polychronicity. The machine remains although the process has stopped. The Gothic here is the coexistence of a machinic structure once sanitized and human-driven with nature. By machinic structure, I mean the structures of control (regulatory mechanisms of the industrial processes) that were organized around protocols designed by humans even when operated by computers or self-regulating processes.

The Gothic, as established by literary fictions and commentators (Punter, Botting), reveled in revenants, ghosts and remnants. Nothing was ever fully dead, finished or interred. Polychronicity in the industrial ruin is the montage of spaces of progress and modernity—of which the factory, for as long as it was running, was the perfect symbol—and nature. Spaces of progress and modernity include the temporal dimension as well, assuming that modernity and modernization indicate a movement toward the future. Nature's reclamation indicates a movement toward the premodern.

We need to situate the UCIL or any other industrial ruin's polychronicity within the campaign for factory revivals. For instance, here is a Reuters report titled "Modi Seeks to Revive India's 'Zombie Factories,' Not Abandon Them," from October 2014:

> The [Modi] government has set up a committee to examine ways sick public companies can be resuscitated, including using cash reserves from profit-earning state firms to provide lifelines to the loss-making ones, according to officials in New Delhi. (http://in.reuters.com/article/2014/10/13/india-industry-revival-modi-idINKCN0I202C20141013. Accessed 25 June 2015)

The report's imagery of sickness, resuscitation and lifelines juxtaposed with the term "zombie factories" merges medical recuperation with voodoo practices. The zombie (originally from the Haitian folklore tradition) is the undead and the staple of much Gothic horror in fiction and film, but somehow seems appropriate here when we recall the mechanized movements of laborers in such factories and the automaton-like movement of zombies. But more than anything else the report captures the Bhopal Gothic's polychronicity by bringing together an (un)dead past, a present and a possible future.

Numerous photographs of the UCIL plant as it stands today are available. Cobweb-laden, rusted machinery strew around fill many of the photographs. (http://www.telegraph.co.uk/news/picturegalleries/worldnews/6711851/Bhopal-disaster-25th-anniversary-of-gas-leak-tragedy-at-Union-Carbide-pesticide-plant.html?image=16 ; http://www.livemint.com/Politics/zyr7dzApQlZOZLnBwVa2XK/Bhopal-gas-tragedy-victims-move-higher-US-court-against-UCC.html, accessed 12 Feb. 2017). In other cases there is lush green growth and ivy winding across and over what was once sophisticated and state-of-the-art machinery (http://www.dailymail.co.uk/indiahome/indianews/article-2156703/Firm-says-damage-Bhopal.html, 12 Feb. 2017).

The ruin is a mnemonic device. On the one hand, it reminds us of a glorious set of fast-moving, rhythmic processes of production, and on the other it reminds us of the potential for decay within all processes. It reminds us of greatness and decay at the same time. The triumphalism of modernity's progress symbolized in the heavy, forbidding buildings and machinery morphs into the silence of decay that has crept up on the same. The ruin's polychronicity induces nostalgia, perhaps for a secular modernity the industry represents even today. It induces sentiment as well, and particularly so in one case.

With Union Carbide, of course, the industrial ruin represents something other than just the dialectic of modernity and decay. The ruin here invokes traumatic memories and, in the wake of environmental activism and the discourses of eco-disaster that surrounds us today, "it is difficult to walk the line between sentimental lament over a loss and the critical reclaiming of a past for the purposes of constructing alternative futures," in Andreas Huyssen's words (2006: 9). UCIL functions as a palimpsest of several narratives, from globalization to neocolonial exploitation, continuing trauma to the differential evaluation of life and death. Polychronicity offers us not just nostalgia but a possible alternative reading of the factory's spaces itself. Tim Edensor writes:

> The topographies of yesteryear thus reassert themselves in memory, the familiar crowd of industrial buildings and the fixtures and local amenities which supported them and their workforce, and in the ghostly traces of past embodied enactions, to produce a sort of phantom network. (2005: 148)

It is not, in the case of UCIL, in the topographies and fixtures that the post-disaster horror lurks. It is in the continuity, the persistence and the polychronicity of flows from 1984 to the present, in the chemical traces running through bodies (water bodies, human bodies) that the derelict ruin asserts itself. The "phantom network" includes the dead, the missing, the victims, the culprits, the absconders. It includes the MIC plants and manufacturing procedures, the dysfunctional safety mechanisms. All of these exist, in ghostly forms, in the UCIL plant's remains but also in the bodies of the people.

There is one more significant mode in which we can discern the Gothicized polychronicity of Bhopal. The UCIL plant is now a thing of the past, but as we have seen, its effects linger in and haunt the bodies and environment of the city. Within the latter generation bodies of the city, UCIL is still alive, just as MIC haunts the genetic materials making up the humans of Bhopal even today. In other words, polychronicity is embedded in the corporeal and environmental ruins of Bhopal. Nature and culture are both subject to ruination here. Tarmo Pikner proposes that even social relations and culture operate in and through metabolizing the "natural" environment. Pikner writes:

> Devices, bodies and (metabolic) ecologies are involved and may become contested in the complex process of transforming some natural entities, such as water for people's use, making nature part of the essential consumption flows of cities. (2014: 84)

In the polychronous scene which is Bhopal the natural bodies of Bhopalis metabolized the chemicals from UCIL, just as the soil and water of the city did. The post-disaster ruin which is UCIL is the effect of a *merged* metabolization, of nature (bodies, soil, air, water) and culture (chemicals from the plant). Just as soil and water are part of the consumption flows that help bodies *live*, the chemical products of UCIL, in and since 1984, became a part of this flow, and determined the metabolisms of bodies, nature and culture. The ruin is the instantiated polychronicity of older chemicals and older metabolization across nature and culture. In the ruins of Bhopal and Bhopalis, we can no longer separate the metabolization flows. The "material vestige" (Edensor 156) of UCIL is not simply in the plant's derelict premises *but in the very bodies of the Bhopalis* with their merged metabolisms.

The polychronicity is itself a reflection of the collapse of teleologies of the modern, or at least one teleology: of industrial modernity.

It could be argued that the postindustrial ruins of Bhopal (and by this I mean not simply the UCIL site but the very bodies, landscape and natural resources of Bhopal) as instantiating in its collapsed teleology of modernity a major critique. This sublime wasteland is produced by the regimes of financial economy and cultural economy of modern times. First, in terms of the financial economy: accumulation, consumption and unregulated production as a way of maximizing profits, minimizing costs, layoffs and waste—all of which we have seen in the case of UCIL's operations. Second, the belief systems—the cultural economy—around production, consumption and cultural protection was never about safety or ethical production. The corrosion of value that leads to waste is located at the concurrence of these two economies, seen in the symbolic representations of wasted landscapes and injured human remnants of postindustrial Bhopal (Nayar 2017).

If one contrasts the postindustrial Bhopal with dystopian sci-fi texts, it is possible to see that the former is not a projection into the future: it is the lingering past and the materially real present. It has been argued that the decadent sublime by extrapolating modernity's conspicuous consumption into the future sets up not simply a *place* where values have collapsed but a *time* when the present modes of valuation generate waste and proliferating waste. If utopia is a *not*-place (etymologically utopia comes from eu-topia, or not-place), and dystopia a *bad* place, then we might think of a *not-when* or uchronia (Sargent 1994) as well. The uchronia represents a fictional time in *our* world. That is, uchronia is our present-day world given to conspicuous consumption, violence and exploitation in a fictional time-period. The decadent sublime doesn't always change the place—earth, the city, suburbia—into dystopia but suggests that this same place might look different in a different time, a no-time rather than a no-place. I suggest that Bhopal is that uchronia: *an unbelievable sci-fi future in the present-day*. If uchronian texts situate a state of affairs that are the *future* culmination of present-day actions in an uncertain tomorrow, Bhopal's polychronicity that captures the decadent sublime of waste, suggests that waste and wasting are uchronian in contemporary India. Waste does not raise a question of *where* (as in wastes will accumulate where?) but when, as in wastes will accumulate *when*: and Bhopal has given us the answer. This is its tragic polychronicity.

THE AESTHETICIZATION OF RUINS

Ruins are increasingly set pieces, settings for commodification, from parties to film shoots. The ruin represents an authentic space of the past, of processes long forgotten or rendered obsolete. But their imbrication with(in) contemporary processes brings the symbols of an older industrial capitalism into consumer capitalism. They are at once authentic ruins and contemporary consumer goods. By filming scenes in such industrial ruins they serve as authentic façades to the whirling signage of the postmodern era.

Andreas Huyssen points out that modernity's obsession with authenticity caused the production of extensive ruin imagery, whether this was in the poetry of Ezra Pound and TS Eliot or in the critical work of Theodor Adorno and Walter Benjamin. The discourse of "authenticity" privileged genius, singularity, creativity, authorship and uniqueness (Huyssen 9). This authenticity was deemed to be under threat from the modern era's powers of infinite reproducibility. Alienation and inauthenticity were at hand, according to the poets and critics. This obsession with authenticity was aligned with the boom in *retro*—from clothing to music to replicas. Huyssen therefore posits the idea of an authentic ruin where the ruin becomes re-packaged, commodified into contemporary consumer products and spectacles.

The industrial ruin is assimilated into the signage of the contemporary. The spectacle of the film enables us to register the ruin without nostalgia but also with no references to its specificity: what kind of factory was this? What factors, political, economic, social, caused its ruin? Did, for example, the structural adjustment program of neo-liberal India have anything to do with the ruination? Ironically, the ahistorical and depoliticized authenticity of this ruination is what enables Bollywood's spectacularly stylized filmmaking and distribution.[2]

Such an absorption into the media world elides historical processes but makes sure the ruin exists in *some* form for consumption as well. The authenticity of ruins merged with the fictional narrative and staging in the film might be perceived as an aestheticization of decay. Filming a ruined factory as part of the general "entertainer" not only revivifies the ruin, it even renders it spectacular. Just as the traditional Gothic aestheticized old castles, forts, labyrinths and decaying houses by making them the sites of lust, hate and strong passions, the contemporary mass entertainment-media industry, aestheticizes the decayed factory. It brings the authenticity of the ruin to the fictional film. UCIL has not been the setting for such films, however.

The industrial ruin is situated between two intersecting economies, which renders it authentic: the ruined economy of the factory and its mechanized production from an older era and the entertainment-media economy of the present. Indeed, the former has been effectively replaced, albeit erratically, by the latter. The fact that governments periodically declare their intentions of reviving the factories suggests that the former economy has not been completely disappeared from the annals either.

Another kind of aestheticization is also at hand in the industrial ruin, as documented in the cultural texts. The industrial ruin on the cusp of being reclaimed by nature returns us to a premodern age, as noted earlier, but it also serves as a space of wilderness in the heart of urban spaces. Bhopal's factory is pushed back in time as a result of such a reclamation:

> No one goes there [the factory], people say it's haunted by those who died. . . . Look inside, you see something strange, a forest is growing. . . . Look throughout this place a silent war is being waged. Mother Nature's trying to take back the land. Wild sandalwood trees have arrived, who knows how, must be their seeds shat by overflying birds. That herb scent, it's ajwain, you catch it drifting in gusts, at such moments, the forest is beautiful, you forget it's haunted.

> Under the poison-house trees are growing up through the pipework. . . . (Sinha 29–31)

Sinha's fictional text paints a portrait of multiple time-zones criss-crossing through the ruined factory, as Nature claims back the land, spirits of the dead haunt the now greener site of UCIL. The only life forms here are plants. Trees grow "through" the pipelines, just as chemical products once flowed through them. Sinha paints a place that is still alive, in a very different sense from UCIL's "life" when it was a working, productive factory.

Nate Millington in an essay on photographs of Detroit's industrial ruins coins the term "post-industrial urban nature" (2013: 285) to describe the wilderness that appears in the abandoned factory. It is not "transcendent wilderness" (284) but "urban weeds" in conjunction with social, economic and political "nature" that has produced this aesthetic setting.

As noted in earlier chapters, all commentaries on Bhopal (Lapierre and Moro, Shastri, Prajapati) make several references to the antiquity of the city, its old palaces and mosques and heritage structures. In the scene, the "messy beauty of dereliction and material decay" (Millington 283) is a mixture of the antiquarian ruins and UCIL's modern ruin. Thus, the polychronicity is also reflected in the layering of ruins between the heritage structure and the derelict plant.

THE WASTES OF MODERNITY AND THE POST-DISASTER UNCANNY

Modernity, argued Zygmunt Bauman (2004), produces waste in the form of effluence but also as wasted people. Avery Gordon has proposed that ghosts and specters represent haunting reminders of modernity's violence. Industrial ruins are odd symbols of a modernity that produced wastes, but also has been rendered waste.

The industrial ruin represents decay at the heart of urban/suburban/rural life. In the midst of activity, buildings, human movement, the industrial ruin is "an absence of stable form and a presence of a void" (Pikner 2014: 85–86). Waste in the case of an industrial ruin is mostly leftover machinery, signage and anything that has not been pilfered. But what is also a reminder, a waste from modernity, is the new object biographies made visible in redundant nonfunctioning machinery.[3]

Considered worthless except in terms of their worth as salable scrap, the machinery and the objects in the factory are *reassigned* value. Instead of the products and objects they once produced for the market they have themselves become objects to be detached from the factory and sold as scrap. Waste here is not an absence of value as much as a *new* value. They do represent *some* economic value as scrap metal, for instance, or cabling/wiring.

Edensor also notes that the ruin makes visible what was hitherto hidden: wires, ducts, pipelines, etc., that controlled the processes when the factory was in production mode (109–10). Like the Gothic in which secrets are revealed as the tale proceeds, the ruined factory reveals its secrets, its processes. Elaborating this Edensor idea along a different tangent, I propose that the exposure of the innards of the factory in its ruined states approximates to the *uncanny* for there is in the uncanny an "apprehension . . . of something that should have remained secret and hidden which has come to light" (Royle 2003: 2).

The uncanny, as Freud points out, relies heavily on doubling and repetition (1971: 386–92). Repetition, in fact, is the very structure of the uncanny. There is also a heavy visual component of the uncanny, which I shall attend to now.

First and foremost, the UCIL ruin brings to the fore an uncanny feeling: this site is at once familiar and unfamiliar. Given the machinery's very real physical presence we recognize it as a site of production, of vast energies, modes of transportation and manufacture of objects. Yet, in the emptiness of the ruined factory we see the familiar defamiliarized: the machinery does not seem to function as machinery should. It is a ghost of a machine, lifeless and therefore valueless. A more socioeconomic rather than merely cultural, dimension of the post-disaster uncanny stems from the ambiguity in the *event* or *place*. Maria Tatar suggests, following the work of Tzvetan Todorov, that the ambiguity in the uncanny event "generates the hesitation that defines the fantastic" (1981: 169). Tatar's insistence on "hesitation" grounds the uncanny in a *worldly* location with attendant contexts of topos, place and perceptions of place. The sight of a particular place or event invokes an uncanny dread because the perceiver hesitates to classify, define and identify the ambiguity in the place or even identify it "properly." Recognizing it as a symbol of power (although now redundant), culture in antagonism to nature, and finally industrial capital that is overrun by plants and odd forms of animal life around still/static machinery the industrial ruin defeats categories of place and space, function and purpose.

As a consequence of the uncanny repetition of the machine in the present, the machinery lies now in the lap of nature. We enter the uncanny in the form of a latent undecidability (Tatar's "hesitation") when we perceive the UCIL grounds and machinery overgrown with bright green plants. The industrial site represents antiquity or the present, since clearly the plant life is of the *present.* Samuel Weber suggests that the uncanny is rooted in undecidability. It implies and involves a second moment or movement: the "defence against this crisis of perception and phenomenality" (1973: 1132). Weber is directing our attention to the epistemological uncertainty that accompanies the troubled perception and cognition engendered by the ruin.

The second instantiation of the uncanny is the coexistence, as noted earlier, of odd objects—such as weeds, grass—in the midst of heavy machinery. Nicholas Royle writes: "uncertainties at the origin concerning colonization and the foreign body, a mixing of what is at once old and long-familiar with is what is strangely 'fresh' and 'new'" mark the uncanny (12). The uncanny "may be construed as a foreign body within oneself, even the experience of oneself as a foreign body" (2). Royle's explication of the uncanny posits a foriegn object within the familiar, so that our perception of the familiar is radically altered. The experience of the foreign as a part of the familiar, however temporarily, is the uncanny.[4]

The co-presence of machinery and weeds, ruins and tourists, the old (ruin) in the new (film) forces us to recognize the incongruity in the familiar, even when the exact nature of that factory is unknowable. As noted earlier, the factory comes back, repeats, is revivified in the form of settings for the contemporary entertainment industry. I propose that the uncanny is also the reassertion of "waste" into the semiotics of contemporary cultural economy. Waste repeats with an entirely different value, purpose and function in this cultural economy, and therefore might be read as the post-disaster uncanny. The icon of modernity turns into its Other: labyrinthine, ghostly, overgrown. But unlike in the traditional Gothic where there is the possibility of rebirth and regeneration, UCIL remains shut down, blackened, darkened and smelly.

Robert Mighall argues: "Gothic space as that which civilisation rejects or demonises" (2007: 61). With UCIL, first welcomed as a savior or even the divine, there is demonization, of course. But it is not a space that civilization has rejected. On the contrary, due to its constant presence in the form of its residues, detritus and pollutants, UCIL remains integral to Bhopal's vision of itself as a community and as a civilization. However, what we can argue about Bhopal's post-industrial Gothic is: UCIL represents for the survivors the demonic at the heart of its civilization, one that it cannot really get rid of or exorcize. This postindustrial Gothic is of the uncanny variety. It still menaces, just as it remains desolate and ruined.

UCIL's ruins in photographs, especially the E610 tank of MIC that exploded on 2/3 December 1984, represents a reprisal of history but also serves as an imaginative geography of both dereliction and decay. Dereliction, because it is the negligence, indifference and corrupting greed that produced the disaster. Decay, because, first, the UCIL plant and its parts had begun to decay but had not been replaced and second, decay is what characterizes the plant and its continuing effect: Bhopal decays due to the ongoing haunting of its Nature by UCIL.

There is another kind of uncanny that asserts its presence decades after the disaster. The ruins of the UCIL plant and the city are in some senses *reanimated*, like ghosts returning, during protests with the performance of the missing persons in the form of metonymic presences: in their photographs (see Colin Toogood's photographs of protestors carrying photographs of their missing family members (http://bhopal.org/torchlight-parade-in-bhopal-marks-30th-anniversary/). The dead are made available, brought back into the public space through these photographs. Marianne Hirsch and Leo Spitzer have argued about family photographs:

> They reveal a normalcy and a social integration that was then violently disrupted and destroyed with the beginnings of persecution, ghettoization, and deportation. Familial descendants might recognize in the photos some of the fabric of family life that had been passed down through stories and behaviors; extra-familial viewers might connect to them in a different way. (2009: 15)

The photographs are uncanny in that they are iterations of a past. They reveal whole Bhopali families and communities. In the post-disaster scenario around UCIL and Bhopal they are indexical signs of former lives, communities and families: they *lived*. The victim photograph is of course a traditional idiom of political protest (the Argentinian Mothers of the Disappeared have, for over four decades, marched carrying photographs of the missing family members). The identity photographs are remainders and reminders of the disruption of families, networks and relationships, and of the failure of the state to protect its citizenry.[5]

The photographs of the disappeared also constitute a regime of meaning that is at odds with the official narrative—even beliefs—about the events of 2/3 Dec. 1984. This regime of meaning captures a set of processes and results—including, of course, death—that do not constitute the officially accepted archive around Bhopal, especially in terms of the number of deaths. The official archive consists of repudiations and denials issued by UCIL (see Hanna et al., APPEN, and other sources), ambiguous legal proceedings of the Indian state or ambivalent expressions of support from the Indian state. Scientific evidence, legal issues and economic explanations are the discursive Gothicization of Bhopal, all of which merely endorse UCIL's, Dow's and global finance's stranglehold. Further, even within eyewitness accounts (written years later) there are photographs of dignitaries visiting the Bhopal affected, reconstruction activities, inauguration of hospitals and processes of "rehabilitation" and compensation. That is, the institutions and organizations find suitable discourses through which their "meaning" (version of events) and disciplinary systems may be enforced, whether these are about denial or about adequate state support. We see this last in Prajapati, for instance, where he captions photographs of relief camps "Safe in Tents" (65), "Opened Coffers for Cash Relief" (55), "Hope Alive" (39) or "Out to Wipe Tears" (55). John Tagg examining photographs from law courts and prisons has argued:

> Institutions whose dissemination constituted the disciplinary regime could only operate, insofar as they did, when certain technologies, techniques, prac-

> tices, and codes of representation had been pulled into place. Integral to the disciplinary apparatus was the production of a specific and novel economy of meaning—a regime of sense that guaranteed a new order of truth and turned on a new structure of documentation whose institutionalized effect was to reverse the political axis of representation, making it no longer a sign of power and prestige to be recorded, but a sign of subjection. (2009: xxxi)

To reverse a regime of truth, itself established and reinforced through these *necessary* "frames" and technologies, as Tagg argues, the protestors and campaigners appropriate the same technologies but resituate it in a new frame and thus seek to establish a counter-hegemonic truth. Prajapati's visual-verbal rhetoric of relief, hope and reconstruction does not fit well with the events that are documented by the campaigners, and such rhetorics is countered by the evidence posted via photographs by the survivors. Bhopal is thus only "knowable" in this contestation of meaning regimes.

UCIL, BHOPAL AND THE ELEGIAC SUBLIME

Years later, commentators like Chouhan and Suman Dey, former employees of UCIL, wander through the ruins of UCIL and explicate the links between this now-defunct plant and the making of deathscapes. They reconstruct the events for us, walking through the ruins, examining bits and pieces of rotting machinery, exploring the grounds around the plant. The vastness of the plant, its intricate machinery, the control room (which must have looked sophisticated and state-of-the-art at some point), the network of pipes, chimneys and valves are embodiments of the technological sublime of the late 20th century. In its present, abandoned state, with the disaster-as-history attached to the technological marvel, it is sublime, but of an entirely different kind, like abandoned industrial towns are in the work of Ambroise Tézenas.[6]

This ruin, like all ruins, elicits and evokes melancholia. This melancholia too is tinged with a sense of the former glories embodied in the factory—the vast machinery, the smooth processes, the large numbers of workers all over the grounds, among others. The films show us the current state of the UCIL control room, once the seat of power and authority. They show us the vast tanks, one of which is E610, that erupted out of the earth and caused the disaster. The network of pipes, the boilers and the chimneys still stand. Reconstructions in CGI or in feature films show us images of a factory in operation as well. Other cultural texts show us present-day Bhopal with the lingering effects of the events of 2/3 Dec. 1984, books provide photographs of the city's mosques and historic monuments, before returning us to the UCIL plant. The link stretching across the city, its many regions and its peoples, is still UCIL.

In the process of reconstruction and recall Chouhan, Dey and others generate multiple emotions: terror of that night, awe at the former glories of the UCIL production system and melancholia at the disaster. The tone in these texts, therefore, is elegiac—for both the city that had been ruined and the factory that deteriorated from its glorious days to first, a source of destruction, and second, a ruin. But the awe around UCIL's halcyon days and the terror around the recall of the events of 2/3 Dec. 1984 are part of this complex of emotions. The result of this messy recall and reconstruction generates what may be termed an elegiac sublime.

First theorized by Michael David Raymond in his study of Wordsworth's *Prelude*, especially Book 13 dealing with the death of John Wordsworth juxtaposed with the image of Mount Snowdon, the elegiac sublime is an aesthetic informed by a struggle to overcome the despair of death, and the massive mountain becomes a symbol of this despair. That is, the awe of the mountain is linked to personal grief and melancholia. It is characterized by "opposing emotions or cognitions joining in the darkness of thought" (222). It is not just a sense of elevated awe at the sight of massive structures—which characterizes the traditional sublime—but is the "perception of dark memory and mist" (222). As Raymond notes, if Tintern Abbey with its "sense sublime" is a symbol of Wordsworth's elevated thoughts, Snowdon represents the "hiding place of his negative elegiac power" (223). It represents "deep transcendent and dark emotions" (228).

Depictions of the ruins of the plant and accounts of the city of Bhopal during and after the events of 2/3 Dec. 1984 embody the elegiac sublime. The progress of the narratives exhibit the emergence of this particular aesthetic. Prajapati in his eyewitness account describes the city *before* the disaster as "beautiful" but also as "over-crowded . . . inhospitable and bleak" (5). From the "ancient Maurya kingdom" it has now "blossomed into a modern city" (13). The leak of the gas "enveloped the entire surrounding city so thickly that the visibility was reduced to a few yards only" (27). Then it becomes a "god-forsaken city" (81). The journalist Anees Chishti describes "a massive ocean of humanity as expansive as the adjacent lake" (25), thus aligning the horror of the leakage, the suffering/dying humanity with the famous lake. Later he visits and documents the "number of graves . . . kept ready for the dead," two days after the events (38–39). He documents anger, fear and anguish among the Bhopalis. Screen shots in *all* documentaries linger over the ruined plant and then range over the abandoned grounds where children play today and the city, where, underneath the apparently bustling life lurks death and suffering in the bones of the survivors and those born to gas-exposed individuals.

The elegiac sublime transforms the awe-inspiring sign of modernity/modernization into a site of melancholic reflection after the disaster. UCIL, like Bhopal itself, is a space in which people once took pride and now can only mourn about. The terrors of that night in 1984 merge with the melancholia of lost lives, lost jobs and collapsed social order. Personal grief becomes symbolized in the ruins. For instance, the anguish over the events of 1984 are writ large on the face of Chouhan when he walks through the UCIL plant in the documentary films. Likewise, Dr. Saxena, whose opinion regarding security and safety measures was resolutely ignored by UCIL authorities, lapses into silence when she recalls the events.

The elegiac sublime is also part of the Gothicization of Bhopal because the ruined UCIL plant and the injured city are demonically twisted versions of the original modern plant and city.[7] It extends before and after the events, and alters perceptions of the space, the people and the process in doing so: the events of 2/3 Dec. 1984 are beyond time and space. UCIL, in the memories of Chauhan, Saxena and Dubey, looms as dark memories, embodying decay and death.

It should be clear that the elegiac sublime in the cultural texts has shades of the horror that echo the Gothic tradition and the terror from the sublime tradition. Contemporary critics on the specific aesthetic category of Gothic sublime in the postmodern age note that such a sublime is marked not by the soaring of the traditional sublime but a "plunging" (Beville 2009: 89, following David Morris's work on the Gothic sublime). Maria Beville discerns, then, the Gothic's "hallucinogenic feeling and the mood of melancholy, mourning anxiety in the text" (130). In encodes and includes "simultaneous fear and fascination, but ultimately, Terror" (153). The sublime itself, as we know, mixes beauty with terror. In Robert Miles's pithy formulation, "A sight of nature's vastness from the top of a mountain would be sublime; the same view from the perspective of someone falling down it would be simple terror" (2007: 14). When Chouhan or Dey view UCIL's remains, or reconstruct the processes for the camera, they view it from the inside, so to speak, from the perspective of one who has seen it all fall apart and fail. As they recreate the plant for us (Suman Dey and Chouhan describe the safety devices, showing us the gauges and the boilers in the plant), the viewers, the beauty of the industry plant, its efficiency and productivity are effaced in favor of the terror, the fear and the melancholic. Mourning the dead plant means mourning the effects it wreaked upon the city. That is, the elegy for the now-defunct UCIL plant is overlaid with the *elegy for the terrifying deaths and continued suffering of thousands*. The sublimity of an industrial marvel is replaced by the elegiac sublimity of mass death and amassed suffering that is Bhopal.

The elegiac sublime in the Gothicization of Bhopal has resonances with these characteristics. While melancholia and the "plunging" feeling is writ large on the faces of the witnesses and those who recall the events of 1984, there is also a fascination we can discern in the way Suman Dey or Chouhan examine the ruins of the UCIL plant. It is terrifying to recall the events of that one night, but it is also, clearly, surprise at having survived and the horror of the mass destruction they experienced around them. The elegiac sublime is terror and awe tinged with enormous melancholy. The journalist Keswani has in fact noted that he must be the most ineffectual journalist ever since he could not stop the disaster from happening. It is this sense of helplessness and mourning that marks the elegiac sublime in the later reconstructions of 2/3 Dec. 1984.

The elegiac sublime is also the aesthetic of collective life-in-death that one can see in the many thousand survivors of the Bhopal disaster. The sheer numbers are themselves terrifying ("I did not know death had undone so many," wrote T. S. Eliot in *The Waste Land*, but here it is life that had undone so many.) But more traumatic is the state of living they occupy. The elegiac sublime is the representation in pamphlets, documentaries and campaign materials such as the *Bhopal Marathon* of the vast numbers of haunted men and women, of expanses of soil and water contaminated beyond repair. It is the collective mourning for a lost city and for entire generations of people who carry traces of MIC in their twisted, suffering and diseased bodies. Visuals of the disused plant, the city of Bhopal and the patients lined up at the Sambhavna clinic that occur in the films draw our attention simultaneously to the sheer magnitude of the plant and its influence, the magnitude of disaster and the interminable nature of suffering. The alignment of the three—the plant, the city and the sick bodies—instantiates the elegiac sublime in the reconstructions.

VULNERABILITY AND THE "PUBLIC INSURGENCIES OF GRIEF"

The events of 2/3 Dec. 1984 and the aftermath, both immediate and continuing, revealed the vulnerability of the Bhopalis. Simultaneously, Warren Anderson's brief arrest, subsequent "flight" and eventual immunity from the legal implications or processes indicate an invulnerability engendered by corporate, legal and state apparatuses. If in the case of the former these apparatuses (UCIL, the Indian state and the judiciary) failed to safeguard the vulnerable Indians, in the case of Anderson/UCIL and Dow, they did. Vulnerability and invulnerability, then, are "politically produced, unequally distributed through and by a differential operation of power" (Butler et al. 2016: 5). Having recognized this, it is now worth examining if, in the thirty years since the disaster, the Bhopalis' vulnerability have been harnessed into something else.

Vulnerability and the Battle against Percepticide

The Bhopal Gothic is marked, as this book has demonstrated, by a haunting by survivors and victims. I now wish to turn to one final aspect of the post-disaster Gothic that is Bhopal by examining the annual assemblies of protesting survivors and campaigners. In Bhopal's post-disaster Gothic, we do not see zombies, werewolves or vampires rampaging through the urban setting seeking blood and meat—the staple of traditional urban vampire or zombie Gothic. In these cases, as Dale Townshend notes, the swarming by the undead and beasts, is a metaphor for proliferating, viral infection (2014: xxvi–xxvii).

Here the crowds of survivor-protestors do not signify a viral threat to the body politic: they expose the autoimmune disaster lurking within the body politic. That is, the crowds are not the infection of the body politic: they instantiate the infection *within* the body politic that generated such injured, haunted bodies. They represent in and through their bodies the *fact* that their bodies have been ignored, and they have been rendered ungrievable. The post-disaster Gothic which is Bhopal is marked by what Judith Butler terms "public insurgencies of grief" (2015: 197) that call attention, through public display, to the ungrievable population.

The assembly, consisting of bodies acting in concert, is a way of "calling into question the inchoate and powerful dimensions of reigning notions of the political" (Butler 2016: 9). "Contestations" of the political, continues Butler, are "enacted by assemblies, strikes and vigils" and by the "bodies [that] are the object of many of the demonstrations" (9). These bodies are "nascent and provisions of popular sovereignty" (16). The body "in its struggle with precarity and persistence is at the heart of so many demonstrations, it is also the body that is on the line . . . enacting, by the embodied form of the gathering, a claim to the political" (18). Further, "when bodies assemble on the street, in the square, or in other public venues is the exercise—one might call it a performative—of the right to appear, a bodily demand for a more livable set of lives" (24–25). The nature of the assembly in Bhopal merits some attention.

The effigies studied in preceding chapters are linked in a performance with human bodies, bodies that are injured, suffering and dying. The bodies assembling around the effigies, cheering, protesting, screaming, crying, are bodies whose lives have been rendered unworthy of life, *zoe* rather than *bíos*. The protest by survivors of 1984, injured, dying, maimed, is an assembly of *broken* bodies. Their claim to the political enacts their banishment from it, and simultaneously underscores how their bodies have been maimed by political decisions. Thus the assembled bodies are at once the embodied effect of political decisions—whether of the UCIL plant's installation, the acquittal of Warren Anderson, the failure of the cultural apparatuses in India, or of the protracted litigation—and the indexical sign of the exclusion from the political realm of "biological citizenship." The bodies are not the "regular" bodies of routine strikes in the case of Bhopal: they are always already political bodies whose very presence in the public spaces of Bhopal's streets signify their exclusion from the public gaze of UCIL, Dow and the Indian state. In other words, they collectively embody precarity. The embodied enactment of precarity in the Bhopal protest assemblies embody a democratic possibility, as Butler argues, but they are also instantiations of the *failure* of the democratic state that had abandoned its citizens.

In other words, the Bhopal protests, thirty years after the events, harness vulnerability into forms of resistance, opening up their injured bodies to the world.

Writing about Don deLillo's "Cold War Gothic" in *Underworld*, Brian McDonald makes the following point regarding crowds in the text: "the crowd gathers with a vague yearning for some sort of shared expiatory experience, for what Mark Osteen identifies as a 'purgative ritual' to militate against that 'unseen something that haunts the day,' and for a broad sense of solidarity" (2008: 104). Bhopal's Gothic is constructed out of not only the eviscerated effigies that we see burning on the streets but the human bodies that have been everted, turned inside out, through diseases, dating from 1984. The yearning for expiation is, in fact, a yearning for being whole again. The collective here functions as an index of the broken *body politic* of Bhopal, of the families, which in turn reflect the broken bodies of *individuals*. Thus, the unseen haunting that McDonald identifies as a key feature of deLillo's Gothic, is the haunting *of* the bodies at the assembly. Haunted bodies haunt the public spaces of Bhopal. Unlike the traditional Gothic, the bodies here do not seek victims: they *are* the victims, they *are* the haunted.

We might therefore think of the post-disaster Gothic that is Bhopal as demonstrating the *assembly of haunted bodies,* highlighting individually and collectively, a shared precarity, of unlivable conditions and unlivable bodies. Their sovereign bodies are no longer sovereign, because these are pharmacogenically kept alive, and their embodiment of what Butler calls "popular sovereignty" is savagely ironic, yet serves to highlight something else. The spectacle of injured, suffering and surviving bodies is an act of political resistance. Exposing the body, when protesting, to, say, police power and brutal reprisals, argues Butler, might serve a political purpose. The Bhopalis—who have been beaten up by the police or state forces for their protests—are "overcoming unwilled conditions of bodily exposure" as a political act (2015: 126–27). Their willing, annual exposure of their (injured) bodies has to do with the exposure of what they unwillingly endured.

We cannot understand, writes Butler, "bodily vulnerability outside of this conception of its constitutive relations to other humans, living processes, and inorganic conditions and vehicles for living" (130). The assembly is about injured people demonstrating their mutual dependency: their shared emphasis on political agency, recognition and rights. In the process, however, they also highlight that what brings them together, the very intent, scope and aspiration driving the assembly, is a historically shared vulnerability. In

1984, the events of 2/3 December exposed their mutual vulnerability: gas, negligence, neglect and abandonment. It is this set of mutually constitutive relations—of shared victimhood and vulnerability—that brought them together, the effect of shared vulnerability unleashed upon them that night.

Then, the mobilization of vulnerability, as Butler terms it, is vulnerability acting "in concert" (151). The injured bodies take over the space that "belongs" to the "normal" human. To return to Butler once more:

> When the bodies of those deemed "disposable" or "ungrievable" assemble in public view . . . they are saying, "we have not slipped quietly into the shadows of public life: we have not become the glaring absence that structures your public life." (152)

It could be read as resistance as well: demonstrating that these injured bodies stake their claim upon the public spaces because these damaged bodies are also the *public*, whose existence has been denied, neglected and subject to invisibilization except in the rhetoric of injury and disaster. The protestors engage in a performative act: they perform their injuries, their identity as injured citizens, their vulnerability that had swerved into helplessness on 2/3 Dec. 1984. If "performativity" means that we are both acted upon and describes "both the processes of being acted on and the conditions and possibilities for acting" (Butler 2016: 18), the Bhopalis willingly make a spectacle of their vulnerability from 2/3 Dec. 1984 but also perform a "deliberate exposure [of this vulnerability] to power" (Butler 2015: 22) today. For Butler this "power of those who are oppressed," expressed in the form of assembled bodies (23) is a sign of political subjectivity and resistance. Butler of course turns subjugation, injury and vulnerability into potentialities for resistance, affirmative identity-making and political action. She is however quick to argue that this identity making and political action cannot be reduced to "free, individual performance" (24), but demands the collective mobilization of vulnerability.

There are three key components to this mobilization of vulnerability.

First, the corporeal activism that drives the mobilization foregrounds injury, incapacity, eroded bodies and persistent medical issues. The bodies present themselves to the cameras in the documentaries, or are presented by others. Foregrounding assorted ailments and complaints, the bodies exhibit an "abject embodiment" (Waskul and van der Riet 2002). Abject embodiment has been defined thus:

> A state in which coherent bodily boundaries erode and the self has little control over the leaking of blood, urine, feces, vomit, bile, pus, and various other hideous body fluids. Its untidiness violates not only biological but also normative boundaries. What is ordinarily inside now comes out, not only threatening

> the concretion of the body but also resulting in an ominous seepage of matter of physical, personal, moral, and social significance. (2002: 487)

The stigma, indignity and suffering experienced by the damaged body is this state of abject embodiment More importantly, as Waskul and van der Riet note in their case studies, the victims/patients experience powerlessness. Yet, the Bhopal campaigns have *foregrounded* such powerless bodies: victims and sufferers have exhibited their abject embodiment. The mobilization of vulnerability is effected and thus "makes the abject subject position explicit" (Jarenski 2010: 98). Shelly Jarenski continues her essay on Ellison's *Invisible Man*: "From this position, which is specifically abject by being 'outside,' the groove of history itself is revealed and the operations of power structures inherent in that groove become visible" (98). The embrace of abjection offers, in this reading, a limited agency to the injured. The foregrounding of abject subjectivity, of powerlessness and location *outside* the medical and legal discourses is an attempt to underscore the power relations that left them outside. Speaking of their exclusions, whether from medicine or the law, and exhibiting their injured bodies they voluntarily mark their exit from the systems that have excluded them. It is an active embracement of exclusion that one sees here, and this, arguing via Jarenski, is not total powerlessness. While Jarenski is partially correct about the *willed* choice of the abject position, it remains uncertain as to whether, after thirty years, the Bhopali electing to foreground voluntary exclusion makes a difference to the larger politics of the campaigns for justice.

Second, the campaigners perform their collective and individual vulnerability by willfully exposing themselves to grievous injury at the hands and instruments of the state apparatus which only serves to highlight that the vulnerable of history are *open to further injury*, and not rescue or help. When the campaigners for justice assembled, over the years, on many occasions they were arrested and beaten up ("30 Years On," 2014). In their attempt to convert their already established vulnerability into resistance, they opened their injured bodies up to further harm, at the hands of the very apparatus designed to provide justice and welfare: the Indian state.

Third, the assumption in these campaigns and representations is that the subaltern's *body* offers the irrefutable truth about the events of the night. Websites that foreground narratives and visuals of assembled bodies in the protests "privilege . . . the subaltern voices as the authentic sites for the production of truth" (Pal and Dutta 2012: 239). This serves as a counter-hegemonic discourse regarding the "truth" of Bhopal to that offered by UCIL or Dow. This assumption pits the material body of the protestor embedded within the discourse of protest against the discourse of UCIL and Dow. Thus, there is an uneven play at work here: of material bodies+discourse versus official discourse.

What is the nature of visibility sought and obtained by injured bodies acting in concert? When the Bhopalis gather in the city of Bhopal or in New Delhi—the seat of the Indian government and therefore the site of most protests—their injured bodies present a certain spectacle. Dwaipayan Banerjee (2013) has made a case for "substance activism" in these protests wherein the campaigners deploy blood, hearts and ketones to gain attention from the Indian state. However, what is crucial about this "corporeal activism" (as Banerjee terms it) is that these bodies share the same "ontological ground and therefore have shared normative scripts" (Hammami 2016: 177) as the rest of the "normal" bodies witnessing their protests. That is, the assembly of injured bodies performs injury on the *same* plane as "normal," uninjured or non-Bhopali bodies who occupy the *same* space as the eyewitnesses to the protests. On this same plane and this same space, the Bhopalis contrast themselves through their corporeal activism from the normative script of their witnesses. The witnesses are forced to admit that these (Bhopali) bodies are *different*, and they are so because of the indifference of the state to *constructions* of this difference. Further, by documenting (as we can see in the photographs of the protests) their embodied protests, they perform a "countersurveillance" and produce "countervisibilities." Those made invisible by the state, whose injuries have been ignored by both UCIL and the Indian state, generate their own visual activism around their own injurd bodies: placing themselves *before* the camera's eye. This countervisibility is the Bhopalis' response and challenge to the erasure by the Indian state of their bodies and injuries.

The post-disaster Gothic is not only the derelict, haunted and abandoned space of UCIL's plant, it is also the horror of this assembly of injured, suffering and ruined bodies in public spaces. While UCIL's plant is sedentary and situated in one specific space, the ruined Bhopali body travels, assembles and generates countervisibilities, whether it is in New Delhi or on the global stage. This embodied assembly by injured Bhopalis with its countervisibility is a response to the "percepticide" of the state and the general "normal" population. Diana Taylor (1997) writing about the "Dirty War" in Argentina and its disappearances argued that "spectacles of violence rendered the population silent, deaf, and blind . . . the self-blinding of the general population—'percepticide'" (122–23). Taylor elaborates:

> [The witness] cannot or will not witness the events taking place in front of her. The military spectacle made people pull back in fear, denial and tacit complicity from the show of force. Therein lay its power. The military violence could have been relatively invisible, as the term "disappearance" suggests. The fact that it wasn't indicates that the population as a whole was the intended target, positioned by means of the spectacle. People had to deny what they saw. . . . To see, without being able to do, disempowers absolutely. But seeing, without

> even admitting that one is seeing, further turns the violence of oneself. Percepticide blinds, maims, kills through the senses. (123–24)

Taylor argues that spectators, by turning away from the sights of state violence around them, collude with the violence.

Taylor is pointing to the willful self-blinding, or percepticide, of and by the witnesses, to the atrocities of the state. In the case of Bhopal, the assembly of injured bodies protests not simply the atrocities of the state—in the form of botched medical treatment, inadequate legal redressal and refusal to clean up the poisoned environment—but the willful blindness of the country to their plight. Percepticide, as Taylor argues, is what enables the state to continue to be indifferent and to perpetuate atrocities. The response to percepticide is countervisibility.

Resilience and Haunting

That the Bhopalis, mobilizing their vulnerability year after year, have demonstrated remarkable resilience in the face of state indifference, state violence, juridical betrayal and continued medical crises may be interpreted, as much of the rhetoric around disasters goes, as a sign of their "spirit." However, the horror of the post-disaster Gothic lies precisely in the *fact* of resilience and the continuing necessity for this resilience in the face of indifference.

Commentators on resilience, such as Sarah Bracke, have argued that the "subject of subaltern resilience" is itself "fetishized by the economic and political institutions that bear great responsibility for the contemporary conditions of precarity" (Bracke 2016: 60. Also Neocleous 2013). Sarah Bracke argues that "resilience . . . is approached as 'raw material' . . . [and] raw material invokes long-standing processes of colonization and international division of labor" (60). That is, it is to the advantage of the global neoliberal regimes that the people of the Global South are resilient because then it enables continuing exploitation:

> Global institutions such as the IMF and the World Bank are currently investing in training programs to build and enhance resilience of individuals, notably those in the Global South, an investment that might be understood as a way to ensure that long-standing and ever-creative strategies of survival are safely molded into the needs of the greedy global economy: a resilient subject is one who can absorb the impact of austerity measures and continue to be productive. (61)

Nicole Shukin makes a similar point about resilience in the wake of Fukushima:

> Although resilience gets fetishized as a resourcefulness inherent to human nature and the ecosystems in which humans are embedded, it demands to be

> critically interrogated as a contingent form of biocapital that is shocked into existence by neoliberal catastrophe, which is to say, catastrophe allowed for and managed by an economic rationality now installed within every sphere of life. Resilience is exploited as a potent resource of flexible labor and life accustomed to the chronically precarious conditions of unlimited growth. If resilient subjectivity is accidentally produced by sudden disaster, it is also consciously cultivated and valorized by corporate and state institutions that have a stake in individuals' and populations' ability to subjectively manage objectively unbearable conditions of life. (2016)

That Bhopalis show resilience in the face of the most obdurate and indifferent legal, economic, social and administrative apparatus is horrific, and not ennobling. It points to the necessity of resilience in order to barely survive, but it also gestures at the indirect *systemic cultivation of this resilience by continually ignoring the survivors.* Further, as we in India have seen in the case of nuclear plants and international agreements around them, the state *expects* the citizenry to be resilient even as it, the state, abandons all responsibilities toward them.

Resilience in the victim populations intersects problematically with the historical rightlessness they have experienced regularly since 1984. Resilience from the victims is a demand made by the very structures that consign them to conditions of rightlessness. Rightlessness has its contemporary sources and origins in "the structural inequalities of market liberalisation and the very functioning of the law" and which makes its presence felt at "many local sites" (Odysseos 2015: 1044). Where global structures of economic and political power ensure that segments of the population—or workforce—are rendered "disposable," it is from these very segments that resilience is expected and demanded. The disposable population, whether the workforce or the victims of an MNC and global company like UCIL, is always local. Note what a victim, Shammu Khan, says about the local-global intersection and the manner in which the injured have been rendered disposable:

> People are still going around in circles for their 1500 rupees relief money. The assets of Carbide are still intact. Neither is the government taking it over, nor is it using Carbide's assets to help the poor victims. The people are not quiet, it's just that they are being lulled. Like when a child cries, one soothes it by diverting its attention saying a tiger is coming or a goat is coming. Neither does the tiger come nor does the goat. And the child eventually sleeps. The government is working in a similar fashion. We will have to cry out all over again. (Hanna et al. 112)

The "living corpse" trope used by survivors to describe their endless suffering and deferred death is a reference to *social* death as well, because they are for the government and UCIL alike, disposable. The victims describe themselves as "zinda laash" (living corpse) in at least one documentary. Another survivor says in an interview:

> We are sobbing, but death doesn't come. Like a living, walking corpse you have left us. . . . If with the strength of your money you had shot us all at once with bullets, then we wouldn't have to die such miserable sobbing deaths. (Hanna et al. 115)

The living dead, especially in the Gothic horror genre and post-millennial fiction, serves as a powerful indictment of the social condition, as commentators have noted. The walking dead constitute, writes Gothic specialist Fred Botting, "images of working masses made redundant and cast aside into living death by postindustrial capitalism's shift from mass production toward consumption and immaterial labor" (2014: 507). Steven Shaviro has argued that the zombie—the living dead—signifies a "vertiginous disidentification that is already latent within our own selves" (1993, cited in Hurley 2007: 145). Expanding Shaviro's argument, Kelly Hurley proposes that the zombie, is a self marked by "indifferentiation, dissolution, blank mindless embodiment, even as it struggles to keep it at bay" (2007: 145), before going to treat this living death as a version of the abject. The trope captures, then, the social death of those who have been rendered "disposable." They occupy a midway point of being alive but not quite, of not being dead but not fully alive either—and they are the direct result of capitalism's most horrific monster-body: UCIL. The "immaterial labor" of postindustrial capitalism that Botting identifies as the context for the living dead is the immaterial bodily matter of the former laborers/workers of UCIL and the entire population around the plant. The living dead of Bhopal are biomedical anomalies, and the end-product of zombification, defined by Travis Linnemann et al. (2014) as the "politico-cultural production" of those who never achieve the status of the living (507). Linnemann et al. add a further layer to the idea of the constructed living dead. Not only are the living dead, like the Bhopal survivors, outside the pale of the state (exemplified in the survivor's cry cited above about how the Indian state has refused to even acknowledge that they exist), but are "active constituents in the fabrication and maintenance of state power" because "the reproduction of police and state power occurs through 'legitimate' violence most often exercised against those dispossessed and 'socially dead' others" (508). The Gothicization of Bhopal produces the living dead through the careful deployment of state (in)action, whether medical, financial or juridical. The survivors-as-living-dead are not, therefore, just the effect of chemical toxins, but the *cumulative effect of the state's processes*. We can summarize in theoretical terms the survivor's cry as follows: "zombification is not merely escapism or entertaining fantasy, but an ideological production that disowns very real, human, material circumstances and consequences" (509). The state having sided *ideologically* with UCIL, Dow and the mechanizations of global capital generates the violence against its citizens. The state's denial of a full life through the rejection of the survivor's right to medical treatment—because of the ignorance around the exact nature of the toxins from 1984—and compensation induces a state of living death in them. First denied full biological citizenship (denial of medical rights has, in fact, been termed a violation of "therapeutic citizenship" for the injured and the sick, see Nguyen 2005) and then doomed to legal, financial and other distress by the state, the Bhopalis *are* the living dead.

There is another interpretation possible of the metaphor of the living dead used by survivors. The survivors believe they inhabit a body that does not quite fit their views as to what their body *should be like*. They are "buried alive in an uncanny body simultaneously familiar and strange," as Becky McLaughlin describes it (2013: 134). Reading the pathological conditions of apotemnophilia (the desire to have one's body parts cut off because the individual feels there is too much of her/his body), McLaughlin demonstrates that the condition's narratives embody a Gothic theme, where the individual experiences her own body as an Other. Haunted by incompleteness precisely because the body is normatively complete, the apotemnophiliac wishes for the body to be revised and altered. It is quite possibly a desire for absence (144). I suggest that the Bhopal survivors, unable to bear the facts of, first and foremost, being alive and secondly, being alive with and in injured bodies very far from the usual bodies, in describing themselves as *zinda laash*, seek a loss of the injured body. They seek to distance themselves from their injured bodies when they employ the metaphor.

The post-disaster Gothicization of Bhopal, with its neglect narrative (embodied in all the interviews and statements of survivors studied by Suroopa Mukherjee, documented in the films, Hanna et al., newsletters and other places) and continual haunting of Nature and people's bodies, serves as a palimpsest through the very resilience of its survivors. The resilience of the Bhopalis ensures the continuity of the precariat public sphere which moves beyond Bhopal to include the entire nation. The documentation of such a neocolonial thanatopolitics, atrocities and consequent large-scale precarity ought to, as studies of atrocity photography have argued about historical events (see Linfield 2010, Twomey 2012), stir up humanitarian responses. The acknowledgement of mutual precarity, Judith Butler believes, will enable such responses. However, as we can see in the case of the postcolony's thanatopolitics, the *reverse* is what happens: vulnerability elicits further violence and exploitation (a response Ann Murphy suggests in her critique of Butler, 2011). This might, at first, seem hyperbolic until we note that the failure of the cultural apparatuses that caused the Bhopal disaster—for example, the absence of proper regulatory frameworks or state monitoring of the plant's safety operations—are discernable in later-day industrial installations and plans. The most recent instance, to contextualize the above argument of Bhopal-as-palimpsest, would be the nuclear deals that India has been signing with the United States, France and Japan. The previous government in India (headed by the Congress—I party) removed all references to the Convention on Supplementary Compensation from the draft of the Civil Liability for Nuclear Damage Act (2010). The 2010 Act had included a provision to hold suppliers (both domestic and foreign vendors of nuclear reactor equipment) indirectly liable for any accident. The Indian government has been under tremendous pressure from the United States and other nuclear suppliers, who have argued that such a clause is detrimental to nuclear technology transfer, to remove the clause. There is also an ongoing tussle over which law would prevail in the eventuality of a disaster on Indian soil, and whether Indian laws would be binding on the foreign supplier. The Modi government, in February 2016, ratified the controversial Convention on Supplementary Compensation for Nuclear Damage, effectively paving the way for nuclear suppliers to provide technology and devices to India.[8] International divisions of labour, the continued protective and uneven trade agreements and resource transfers that are the hallmarks of globalization do not allow either actual or normative operations of national culture or national sovereignty.

The worrying aspect of this thanatopolitics that merges neoliberal agendas with an indifference to imminent hazards and biomedical disasters, of which Bhopal was the first, is that there is an overreliance on the resilience of human bodies *in the face of intransigent objects*.[9] That is, where MIC is the intransigent, non-neutralizable object (in fact, as argued, it is a hyperobject) whose tenacity and persistence cannot be altered, the human body is called upon to develop and exhibit resilience. This demand upon the human body in the face of object-intransigence is also made in the context of corporate intransigence, such as that of Dow, whose refusal to accept any responsibility for UCIL or Bhopal has remained unwavering. As Michael Parker, President and CEO of the Dow Chemical Company, stated in a letter to Dow employees on 28 Nov. 2002:

> What we cannot and will not do—no matter where Greenpeace takes their protests and how much they seek to undermine Dow's reputation with the general public—is accept responsibility for the Bhopal accident. (Hanna et al. 248)

Bhopal's precariat, while demonstrating resilience, has *remained* the precariat. Their resilience has been instrumentalized in order to survive and fight for justice, but their precarious lives are instances of continuing political subjectivity at the hands of the state and the corporates. Just as the Bhopalis were the "raw material" (as Sarah Brucke characterizes them) for UCIL, all Indians are potentially raw material for nuclear, chemical and other industries that are being welcomed into the country. In other words, to praise the continuing resilience of Bhopalis is to foreclose the possibility of imagining safeguards, regulatory frameworks and healthy cultural apparatuses. To foreground resilience is to do away with the alternative ways of doing things, such as liability clauses or safeguards, relying as it does, all the time, on the *Third World native to train himself/herself to be resilient.* The post-disaster Gothic is marked by a horrific scenario: the only agency possessed by the politically constituted precarious subjects is the agency to develop resilience to imminent, potential disaster and injury; for the social order, the legal system and the state will *not* transform themselves or work in alternate ways. The rhetoric of resilience, in short, ensures that the Bhopali is any Indian in the neocolonial age.

MATERIALITIES AND POSTCOLONIAL THANATOPOLITICS

Bhopal, in David Arnold's words, "re-enacted, on a far greater and more violent scale, many of the themes—of insufficiently regulated industrial and urban toxicity, of management negligence and profit seeking above safety, of poor governance and even poverty—that had been evident in India's toxic history for decades" (2016: 205). But what has been worrying, as Arnold points out, is that the "catastrophe had some, if only temporary impact on policy in India" (208). The emphasis on "global economic status and increased industrial productivity" in India's recent past, Arnold suggests, continues to make possible toxicity current (208). Arnold is pointing to a toxic history that had, once, climaxed in and with Bhopal but which remains an informing presence in terms of lax environmental laws and adequate safety nets for the people. This toxic history, which encodes neglect and indifference, political and economic profiteering, is India's current thanatopolitics as well, as this book has hopefully demonstrated. A toxic history, present and future, after Bhopal, must be seen in terms of the *material.*

The problematic political ecology (as Bruno Latour 2004, Jane Bennett 2010 and others term the dynamic relationship between the human and nonhuman) linking intransigent objects (chemicals such as MIC), environments, corporate bodies and human bodies places the onus of recovery on the humans, rendering them the precariat because there are no longer sustaining environments in which they may live. Bhopal, as the conclusion to the third chapter discussed through Indira Jaisingh's statements on the need to safeguard India's sovereignty in the light of the disaster, the intransigence of the material and the corporate bodies that produce the material, demonstrated the dangers of socially constructed disasters.

When hyperobjects such as MIC possess the kind of agency that Bhopal demonstrated they possess, then the questions before the postcolony are:

what are the social formations accruing or growing around the agential actants such as toxins? If we assume that non-human objects such as MIC have agency and are actants in social formations, then what is the nature of the politics of the postcolony that enables these actants to perform their functions? An emphasis on hyperobjects and agential objects, such as chemicals, must necessarily turn the discourse of haunting and continued disaster to materialities and the postcolony. It should be evident, from the study of haunting in the two chapters that Bhopal is haunted by material objects—toxic chemical residues in the soil, in the human DNA and in the water. Haunting therefore is the persistence of invisible but palpable *matter*. Bhopal forces us to acknowledge the agential role of supposedly inert matter. Rethinking "matter" in "the new materialisms," Diane Coole and Samantha Frost write:

> If everything is material in as much as it is composed of physicochemical processes, nothing is reducible to such processes, at least as conventionally understood. For materiality is always something more than "mere" matter: an excess, force, vitality, relationality, or difference that renders matter active, self-creative, productive, unpredictable. In sum, new materialists are rediscovering a materiality that materializes, evincing immanent modes of self-transformation that compel us to think of causation in far more complex terms; to recognize that phenomena are caught in a multitude of interlocking systems and forces and to consider anew the location and nature of capacities for agency. (2010: 9)

Matter such as cyanide, phosgene or any of the chemicals (studies conducted in Bhopal's soil and water have revealed these) are part of a web of processes, from the moment of their production within UCIL to their sedimentation in the environment (see, for instance, the Madhya Pradesh Public Heath Engineering Department's Report on the Presence of Chemicals in the Ground Water in the Vicinity of the Union Carbide Factory, 1996; the National Environmental Engineering Research Institute report, 1997; the Madhya Pradesh Pollution Control Board Ground Water Analysis, 2004, among others, extracted in Hanna et al., edited *The Bhopal Reader*). Their effect, asserted in the DNA, the bones and tissues of humans reveals both self-transformation and agential role, especially when they come in contact with live or organic tissue. Thus, what a postcolony with Bhopal in it needs to understand is that matter is caught in a multitude of interlocking systems and forces, which includes the *social* and the *political*. The postcolony as represented in Bhopal, thirty years after the disaster, exemplifies these systems where the biopolitical economy of bodies is informed and mediated by the inorganic matter within them. The biography of a Bhopali is now, effectively, the biography of the matter within.

In terms of the postcolony's thanatopolitics, then, we need to think in terms of the *materials*—chemicals and bodies, industrial machinery and documentary evidence—produced at the nexus of a globalizing economic-political policies, the industry and organic bodies. (The protest campaigns have attracted global participation, from signature petitions to hunger strikes and sit-ins. See Pal and Dutta 2012.) Indeed, Bhopal, along with Chernobyl and Fukushima is one in a series of "neoliberal catastrophes," as Nicole Shukin calls it (2016). Bhopal's bodies represent ready *biocapital* for the neoliberal agenda, and thus become the victims of the toxic material instantiation of this agenda as well. Their bodies are interlocked with the matter produced in UCIL and such "bodies." In the age of globalization, and the dominance of virtual money and spectral capital, material bodies engaged in material labor have not disappeared (as Hardt and Negri pointed out, 2000). The labor inside UCIL is supplied from Bhopal city with ancillary support and services (such as the tea-vendor who would be an eyewitness about the situation in the control room that crucial night) from the neighborhood. In the neoliberal catastrophe scenario, especially one where the catastrophe is ongoing, the materiality of bodies and the social *bíos* is rendered into *zoe*, bare life, or barely living bodies (Giorgio Agamben's categorization of life as socially and politically relevant life, *bíos*, and mere living, or *zoe*, 1998). Their disenfranchisement and invisibility from the circuits of global biocapital ever since they became injured—itself due to their role as social *bíos* in the MNC-driven economy of the neoliberal sort—suggests that their injured materiality has ceased to *matter*. That is, their material bodies mattered as long as they were a part of the interlocked human bodies'—corporate bodies'—global capital network. Their role in the circulation of materials positioned them as critical nodes in the neoliberal scheme, possessing biopolitical *value*. Injury which renders them *zoe*, and the concomitant demands for justice, however relocate them to the domain of immaterial labor: bodies that do not matter (anymore). The thanatopolitics ensures that, to adapt Achille Mbembe (2003), decisions about life made by the state in the case of the Bhopalis are effectively decisions about their death. While death is neither engineered directly nor is immediate, the state of the living dead that we have discussed above suggests the ideological construction of biopolitically marginalized bodies.

When the protestors display their injured, deformed, sick bodies in their posters and campaigns these visuals work to *render* the material violence of the nation's policies and systemic flaws. I adapt the term "render" from Nicole Shukin's perceptive reading of animal dissection and biocapitalization in her *Animal Capital* (2006). Shukin writes:

> *Rendering* signifies both the mimetic act of making a copy, that is, reproducing or interpreting an object in linguistic, painterly, musical, filmic, or other

> media (new technologies of 3-D digital animation are, for instance, called "renderers") *and* the industrial boiling down and recycling of animal remains. (20)

Shukin is interested in exposing the "complicity of 'the arts' and 'industry' in the conditions of possibility of capitalism" in the form of the animal body (20). Shukin continues:

> An interimplication of representational and economic logics is pivotal to biopolitical critique, since biopower never operates solely through the power to reproduce life literally, via the biological capital of the specimen or species, nor does it operate solely through the power to reproduce it figuratively via the symbolic capital of the animal sign, but instead operates through the power to hegemonize both the meaning and matter of life. (20)

While Shukin mounts her critique of animal capital through a reading of this "interimplication," it is possible to see how a critique of biocapital might also use a similar strategy.

In the winter 2010 issue of *777*, the *Bhopal Medical Appeal Newsletter*, we see a photograph of a boy holding up his hands, with fused and bent fingers. The photograph is a rendering—of an injured body, of a body that was not present at the time of the disaster but has inherited toxic *matter* through the genes of the gas-exposed parent(s). The rendering of the injured body makes visible and embodies matter that lies too deep for rendering in and of itself. The *zoe* that this body is, or has become, therefore, reverses the biopolitical equation by being rendered into political discourse. That is, where UCIL and the subsequent politics ensured that *bíos* became *zoe*, the campaigns by rendering visible the injuries in various forms, once again repositions the body at the heart of the struggle. There is, clearly, another form of the biopolitical that emerges out of the protest-politics of the injured material body and its rendering.

When the protests juxtaposed the injured bodies alongside their demand for bringing Warren Anderson home, or associating Anderson with terrorism just as the United States did with Osama bin Laden, they again highlight the implication of the local crisis within global dynamics (as argued elsewhere, Nayar 2014). The "Inhuman" campaign from the Bhopal protestors, archived at www.bhopal.net, responding to Dow's Human Element campaign of 2006–2009 (https://www.youtube.com/watch?v=vsCG26886w8), foregrounded the postcolonial recognition of the global threat such corporations and their processes posed.

Dow set the video in a wide variety of places, from Asia to the USA, deserts to tropical forests, suggesting that in any landscape, the common element that binds is the Human Element. The Human Element is foundational to all matter on earth and is the catalyst that energizes all processes on earth. Dow brings together multiple landscapes, races and cultures into the ambit of its feel-good video. Common to the world, every corner and every topographical feature, suggests the infomercial, is the Human Element. Dow emphasizes the *universality* of the human spirit and the human element itself. The Bhopal protest campaign shifted its emphasis when responding to this "universality" theme: Dow is a universal killer.

In one of its "Where Is the Human Element" campaign posters, the image of an old, wizened woman is accompanied by a list in the text: Agent Orange (Vietnam), Nemagon (Nicaragua) and Bhopal. The universal in the campaign is the exposure to biomedical disorders of millions of people worldwide in what is a truly horrific version of globalization. The Bhopal campaign does not speak of universalism except as *universal suffering* and *universal vulnerability*. Another poster proclaims: "every single person on Earth carries some of Dow's poison in their body." If the Human Element in Dow's infomercial is universal, then toxins produced by the company are also universal. The campaign suggests that, from Michigan to Saigon, we are well en-Dowed humans united through a shared toxicology report. Dow's global reach has little to do with either its safety or progress (some would recall the savage irony of the sign board in the Carbide factory, Bhopal: "safety is everybody's business"—which every documentary focuses on). Every place on earth, in other words, is a Bhopal. Or, Bhopal is everyplace that Dow is. The singularity of a disaster such as Bhopal 1984 is embedded in a universality of imminent catastrophe. Every disaster might be singular in its specific detail, but the processes that culminated in the disaster are not singular in themselves. The greed, neglect of safety features, carelessness, culpability and deliberate production of potentially lethal toxins is common to any of the cities and provinces with a Dow or Carbide installation.

The thanatopolitics of the postcolony, then, is a national failure (in terms of the Indian state's inability to ensure justice for Bhopalis), but one which cannot be delinked from international biopolitical regimes.[10] This thanatopolitics which denies even basic therapeutic citizenship to the injured is also a denial of both, the state's sovereignty and its global power.

If the embedded poisons pose one set of problems then the absence of fuller safeguards in terms of both liability policies and operational mechanisms in the postcolony's neoliberal phase pose another. If Bhopal was always imminent, as the first chapter's examination of texts revealed, then, the possibility of "other Bhopals" remains a clear and present danger (hence the slogan of the protestors, "No More Bhopals"). Thus, the disaster that is Bhopal is also frightening, in true Gothic fashion, for its iterability given the structural conditions of local-global capital movements, industrial and environmental policy in which such plants continue to operate. Commentators probing disasters in filmic representation have argued that "visualizing the potential for disaster in social structures via a representation of the vulnerability of the body and the structures in which it is embedded is really a concern with the reproducibility of disaster" and because "biological disaster is reproducible owing to the capacity for destruction always already embedded within our social and cultural structures" (Sundaram 2012: 137).

The thanatopolitics that produced Bhopal and continues to inform and mediate the biomedical and political bodies of the Bhopalis but also its very land and water marks out a space within the sovereign state that is India which is, oddly, UCIL still. This thanatopolitical space demarcated by the hyperobject that is MIC is intergenerational. A thanatopolitics found in the flawed political ecology of neoliberal policies that allow foreign (corporate) bodies and objects to operate and reproduce has gathered strength since 1984. To be rendered abject in 1984 is to pass on or inherit abjection in the thanatopolitics of the postcolony.

NOTES

1. See the data at https://data.gov.in/catalog/category-wise-wastelands-india#web_catalog_tabs_block_10).

2. In recent times in India we have seen the industrial ruin figure in successful Hindi films such as *Dabangg* and *Rowdy Rathore*, and in highly stylized and expensive ones like Sankar's *I* (Tamil's biggest grosser of 2015).

3. Tim Edensor writes: "Since factories produce things for markets, objects situated in ruins also thereby escape another kind of classification in that they join the realm of waste products and are removed from the ranks of separate artefacts on shelves, in job lots, in typologies—according to function and price—that present themselves as Commodities" (100).

4. In the case of Union Carbide (as in the case of other similar industrial ruins) various myths, stories and biographies circulate, lending a specifically uncanny air to the ruin. Ghost stories set in ruins are part of the postindustrial uncanny for, as Hélène Cixous argues the uncanny is connected, to a series of anecdotal examples, literary and biographical mini stories, its "mythic anthropology" (1976: 539). What we see here, then, is a wholly new biography of the industry emerging, radically disconnected from the past, but appropriating the archaeology of the place in order to offer an uncanny, recuperative, horrific (as in the rape case at Shakti Mills, 2013) or even redemptive narratives.

5. The role of photographs of victims and disaster in the development of distant sympathy and spectatorial humanitarianism, dating back to the 19th century, has been discussed extensively. See, for instance, Fehrenbach and Rodogno (2015).

6. See, for instance, his collection in "dark tourism," *I Was Here* (2015).

7. Tabish Khair has argued: "the God of Gothic literature retained some sublime features, if only in 'demonic' forms distorted into a mirror image of the divine" (2009: 63). Further, such a sublime, writes Khair, is "something vast, non-rational, non-human, something always exceeding the 'now' and the 'here'" (67). The elegiac sublime is akin to such a sublime in the Gothicization of Bhopal.

8. The Ministry of External Affairs, on the FAQs section of its website (updated Feb 2015), claims there is "no proposal to amend the Act or the Rules" (http://www.mea.gov.in/press-releases.htm?dtl/24766/Frequently_Asked_Questions_and_Answers_on_Civil_Liability_for_Nuclear_Damage_Act_2010_and_related_issues. 16 Dec. 2016). However, the supplier nations have been voicing "concerns" about the clauses that are likely to hold the supplier responsible (see http://www.ndtv.com/india-news/concern-over-indias-nuclear-liability-law-still-remains-french-firm-edf-1398896).

9. Recent work on thanatopolitics suggests that it is no longer about the life and/or death of populations and individual bodies. Rather, argues François Debrix (2017), the horror of con-

temporary thanatopolitics is to render human bodies, *beyond* the fact of life and death, unrecognizable, even undistinguishable from non-human matter.

10. Christopher Jones et al. have critiqued the representation of the Fukushima disaster as a "national failure": "[This] focus on events at the national scale underplays local and international causes and effects. The dominant storyline emphasizes corrupt state and industry officials without situating them in the international matrix of forces that created the Japanese nuclear power industry" (2013: 603).

Conclusion

"Burial of an Unknown Child" as Icon

Numerous photographs exist from the Bhopal tragedy. Yet, if we were to consider what constitutes the single most iconic image that recalls the disaster, it would have to be Raghu Rai's photograph, "Burial of an Unknown Child" (Rai, *Exposure*, www.magnumphotos.com).

By way of a conclusion to a tragedy that has *no* conclusion, since the survivors continue to suffer from the horrific effects of the gas, I examine the iconic status of this photograph, and why it means so much to us.

If one were to contextualize the photograph in the cultural texts of Bhopal that this book has studied, then, one of the first frames would be that of the slums and poverty of the UCIL and its neighborhoods. All films linger over these slums, as already noted. While not exactly akin to "slum tourism," "dark tourism" and "poverty porn," the Bhopal environs are inserted into the visual economy of the contemporary. Uli Linke refers to signs of "urban poverty [that] are imagined, trafficked, branded, and consumed along transnational media circuits as signifying practices in a global panopticon of race" (2013: 1224). Linke proposes that "slum iconicities are not merely consumed as cultural exemplars or artifacts of non-European urban worlds but also interpreted as visual evidence of racial alterities" (1224). India and Bhopal represent in such photographs and visuals a certain "embedded injustice" (Ewalt 2011: 339–41) as a part of its very geography.

While this argument about a certain "tropicalism" ("representational practices that belong to those structures of domination that relegate entire populations [of the global South] to the economic and social margins of urban space," Linke 1225–26) is not to reject the global humanitarian efforts to help Bhopal's survivors, the visual rhetoric of the cultural texts does suggest

“Burial of an Unknown Child” by Raghu Rai.

an alterity that is racially, nationally and geographically coded. It is the sheer otherness of Bhopal's contexts—amplified by the cultural texts' insistence on how safety measures and policies were radically different in UCIL plants in the United States and India—that frames the photograph. That is, Bhopal's previvors and poverty are slum iconicities circulating in global media circuits of suffering and aid as the Other face of industrial modernity: the photograph is framed within this set of iconicities.

There is a certain tenderness with which the hand brushes the mud and gravel around the child's face/head. The tenderness, standing as a sharp contrast to the horrors of 2/3 Dec. 1984, is gut wrenching because it also signals a *delicacy*, a *propriety*, toward the dead child which the child was possibly never accorded in the moments of dying. There is, in the act captured in the photograph, a fantasy or illusion of decorum which Bhopal did not possess or exhibit in the course of the disaster, nor was it allowed the victims and survivors. However, the bleached staring eyes of the dead child, the anonymity of identity sits oddly with the delicacy and propriety of the brushing hand. I suggest that what arrests us is this very conjunction of the delicate and proper with the grotesque. Following the work of Yael Shapira reading the "delicate Gothic," I suggest that Rai's photograph brings into unassimilable convergence "a radically purified ideal and a scandalous, spectacular grotesque" (2006: 9). The image of the respected, dignified body being treated with care is an odd reminder of the sheer grotesque nature of the deaths in Bhopal (everted physiologies, choking and coughing to death).

Further, the act of burial, signifying the hiding away of (one kind of) evidence of UCIL and Bhopal's disaster *breaches* the child's dignity in death. It exposes the dead endlessly in the circulation of Rai's image, even as the body itself is "shot" in the act of burial.

Images become icons when they fulfill certain specific conditions. Martin Kemp in his *Christ to Coke: How Image Becomes Icon* defines a visual icon as "one that has achieved wholly exceptional levels of widespread recognizability and has come to carry a rich series of varied associations for very large numbers of people across time and cultures, such that it has to a greater or lesser degree transgressed the parameters of its initial making, function, context, and meaning" (2012: 3). The associations that Rai's photograph could be varied, except for one specific state: this is the burial of an *infant*, and represents a life cut drastically short.

Rai's photograph is an instance of what historical biologists term an "extinction icon" (Turvey and Cheke 2008). The burial of the child marks the extinction of entire families, generations and communities of previvors—Bhopalis—due to circumstances over which they had no control. If, as Turvey and Cheke examine in the case of species extinction, the human causes of such extinction could be diverse (and disputed), Rai's photograph read in conjunction with the evidentiary documents of Bhopal becomes iconic for cathecting the entire faulty technological, administrative and cultural processes.

Rai also captures the ultimate and the ideal victim in the photograph: the child. Nils Christie (1986) proposed the following features of the "ideal victim": the victim is weak in relation to the offender; the victim is, if not acting virtuously, then at least going about their legitimate, ordinary everyday business; the victim is blameless for what happened; the victim is unrelated to and does not know the "stranger" who has committed the offence; the offender is unambiguously big and bad; the victim has the right combination of power, influence or sympathy to successfully elicit victim status without threatening (and thus risking opposition from) strong countervailing vested interests.

Victimhood is a form of communication, where the injured party needs to communicate the experience, argues Rainer Strobl (2010). For this communicative act to be meaningful, it requires a set of norms and cultural frames that bestows a special social status upon the injured. A person might regard herself/himself as a victim and the world agrees with this; a person may not regard himself/herself as a victim and is not regarded as a victim by relevant others; a person could regard himself/herself as a victim but not be regarded as a victim by relevant others; a person could not regard himself/herself as a victim but be regarded as a victim by relevant others (Strobl).

This is precisely why the child victim draws our attention: the child victim does not have a language to communicate in. The child victim communicates not by establishing claims but simply by being a child. The martyr as witness, we have been told by Jacques Derrida, does not always offer his story: he offers himself, his body as testament to his faith. This is the body of suffering that communicates as a body whose disintegrated state immediately warns of the loss of corporeal coherence in a being/person who has not yet attained corporeal coherence in any wholeness. This spectacle of the child victim is a version of horror fiction and film where the dissolving body is a staple image, writes Kelly Hurley (1994), with the difference being: the body of the child victim has not dissolved or been dismembered (though there are those kinds of images as well), but is always positioned at the cusp of such a dissolution. It is not the disintegration as much as the processes and moments leading up to the imminent, even unstoppable, dissolution that horrifies.

The child victim of specific crimes—abuse, murder, etc.—occupies a separate space. Here the victim (for instance the victims of Ian Brady and Myra Hindley, or Jeffrey Dahmer) is, or had been, weaker than the perpetrator or offender. The victim attains victim status through this unevenness of agency and power. There is an individual who might be held to be guilty and who, theoretically, may be brought to justice for harming the person. Such victims are radically different from Phan Thị Kim Phúc OOnt (the Napalm girl in Nick Ut's Vietnam photograph) or Aylan Kurdi (the Turkish boy drowned as they sought to escape the IS) because, in their case, there is no single, identifiable individual perpetrator. Undoubtedly, child victims of serial killers attract sympathy and outrage which then is channeled against the perpetrators (if caught, as illustrated in the social outrage against Brady and Hindley). Thus there is a clear victim/perpetrator binary at work in such cases.

In the case of Phan Thị Kim Phúc OOnt or Kurdi or Rai's "unknown child," the moral status of the victim is amplified because one cannot find an individual perpetrator and only a perpetrator-situation. War, riots, natural or man-made disaster are perpetrator situations in which the child victims are produced and nobody can be brought to justice for their victimhood. It is the impossibility of ever acquiring justice for the child victim that makes Phan Thị Kim Phúc OOnt or Omran Daqneesh unique in terms of child victim status. These child victims are victims of the world at large, of systemic conditions and events.

The child victim is an ideal victim because s/he walks into, or is at, the "scope." S/he is at once the site of "skopos" and "telos." Samuel Weber (2005) makes the distinction between *skopos* and *telos*, where the former is the target and the latter the fulfillment of an action or set of processes: "Skopos is already, tendentially, the tele-scope, since 'the one who aims' is also 'the one who surveys.'" To survey, in this sense, is to command at a distance. Skopos then "designates not just the act but also the object of such watching: the mark or target." In wartime, the imminent victims are watched and then they walk into the firing range (Phan Thị Kim Phúc OOnt, Rai's "unknown child"). The child's victimhood results from being a target of opportunity, whether in wartime, natural disaster or industrial disaster. The child simply happens to be in a zone devoid of any resources for help or survival. S/he is scoped for suffering and/or death, so to speak. S/he is not an individual target, but is nevertheless the individual face of the event.

The image of the child victim becomes an icon precisely because there is no visible and identifiable perpetrator, no obvious reason for the child to be caught in the cross-fire or to die in a blaze of gas. Raghu Rai's "unknown child" does not need a context, or a frame: indeed, it *is* the frame through which we see the needless, casual deaths of Bhopal. The child is scoped into death.

Rai's iconic photograph defines death, because the child is dead. There is, also, a certain compositional unity in the photograph. As the child lies dead, a hand carefully arranges the soil around it, perhaps leaving the face open till the last. The strangely translucent eyes of the child frozen in death is in sharp contrast to the dynamics of the hand and soil around it. It is this binary of permanent stillness and moving hand ensuring a safe resting spot that engages us, and renders the photograph iconic.

Martin Kemp argues that icons should be at once specific and yet not so. It is at once both of the moment and not of it (219). Rai's photograph could "belong" to any age, to any disaster. It at once is the prototypical victim photograph and yet remains deeply contextualized as *the* Bhopal photograph. Adapting work from celebrity studies, I argue that Rai's photograph organizes an entire social imaginary around it (as Celia Lury has argued about icons, 2012), an imaginary that draws attention to needless death, the anonymity of death and the absence of a future to that single child, although the child itself represents the aspirations of several. There are two discourses that intersect in the form of the photograph: the discourse of identity/identification and the discourse of death. The absence of markers renders the death of the child more poignant because it forces one to consider the ones who did not die with the child, and who might have looked for this one child. The anonymity of dying here seems to render the child, any child or every child. The second discourse, of death, is instantiated in the very act of burial, and the structuration of the process where the rest of the body has been more or less concealed with the face alone left open, for the moment. The discourse of identity intersects with the discourse of death in this brief moment before the final concealment of (even) the face, almost as though somebody waits for a last-minute identification of the corpse, a revelation of the true identity of the child.

Such a photograph originating in the Bhopal disaster not only becomes a symbol for the "unimaginable" but also structures our perceptions and receptivity to new disasters and new tragedies (Brink 2000: 135–36), or what we can think of as a social imaginary. The iteration of Rai's photograph over the years is, I argue, the potential instrument of such reformation of the social imaginary. Iconicity here then is not about the death of *one* child, but what one death—and how it resulted—has *come to stand in for* and what it calls upon us to contemplate: systemic failure, precarious lives and lifelong ecological disaster. Rai's "unknown child" *represents* something *other than* itself, something more than itself. Its cultural iconicity lies in the photograph's ability to serve as a template for any and every tragedy, if one so chooses. Elsewhere brand studies scholar Celia Lury has argued that the icon enables us to make "abductive inferences about the possibilities of objects" (2004: 69). That is, when we reference an icon we immediately call attention to various qualities of the icon but also of all those objects with whom we connect the icon. That is, when an icon is transposed (like Che Tee-shirts, Shakespearean adaptations) or "abducted" to other objects, we infer *possibilities* for those objects as well. Rai's photograph transposed into any context forces us, like any iconic photograph, to pay attention to qualities, processes and events that *could* produce similar conditions of child victims.

Celia Lury later argues that an icon becomes a source for the elaboration of "the social imaginary or ground of abstraction as the possible field from which relations of similitude might be inferred" (2012: 254). If so, Rai's photograph forces us to generate a social imaginary of disaster, innocent victimhood, corporate greed and state indifference. It calls upon us to imagine potential disasters that generate such victims as the "unknown child."

The campaign cry for Justice for Bhopal says, "No More Bhopals," a cry reiterated every year. This is a verbal equivalent of the *continuity* of movement of Rai's photograph across time and space. The Rai photograph in its very act of iteration *cautions against the iterations of such events warranting similar photographs*: this is what makes Rai's photograph iconic.

Bibliography

25 Years after Midnight. Documentary. CNN. 2009.

"30 years on, fight for justice still rages in Bhopal," *Times of India* 30 Nov. 2014. http://timesofindia.indiatimes.com/india/Bai-was-living-just-outside-the-Union-Carbide-factory-when-around-30000-tonnes-of-methyl-isocyanate-MIC-began-spewing-from-a-tank-shortly-before-midnight-/articleshow/45323030.cms. 16 Dec. 2016.

777: The Newsletter of the Bhopal Medical Appeal.

Agamben, Giorgio. *Homo Sacer: Sovereign Power and Bare Life*. Trans. Daniel Heller-Roazen. Stanford: Stanford UP, 1998.

Alexander, Jeffrey C., Ron Eyerman, Bernard Giesen, Neil J. Smelser, Piotr Sztompka. *Cultural Trauma and Collective Identity*. Berkeley and London: U of California P, 2004.

Alexander, Jeffrey C. *Trauma: A Social Theory*. Cambridge: Polity, 2012.

Amnesty International. *30 Years Is Too Long . . . to Wait for Justice*. London: Amnesty International, 2014.

———. *Clouds of Injustice: Bhopal Disaster 20 Years On*. London: Amnesty International, 2004.

Anolik, Ruth Bienstock. "Introduction: Diagnosing Demons: Creating and Disabling the Discourse of Difference in the Gothic Text," in Anolik (ed.) *Demons of the Body and Mind: Essays on Disability in Gothic Literature*. Jefferson, North Carolina and London: McFarland, 2010. 1–19.

APPEN (Asia-Pacific People's Environment Network) *The Bhopal Tragedy—One Year After*. Penang: APPEN-Shahabat Alam Malaysia, 1985.

Arendt, Hannah. *The Human Condition*. Chicago and London: Chicago UP, 1998.

Arnold, David. *Toxic Histories: Poison and Pollution in Modern India*. New Delhi: Cambridge UP, 2016.

Arya, Rina. *Abjection and Representation: An Exploration of Abjection in the Visual Arts, Film and Literature*. London: Palgrave-Macmillan, 2014.

Baldick, Chris. "Introduction," in Baldick (ed.) *The Oxford Book of Gothic Tales*. New York: Oxford UP, 1992.

Banerjee, Dwaipayan. "Writing the Disaster: Substance Activism after Bhopal,"*Contemporary South Asia* 21.3 (2013): 230–242.

Baudrillard, Jean. *The Illusion of the End*. Cambridge: Polity, 1994.

Bauman, Zygmunt. *Modernity and the Holocaust*. 1989. Ithaca: Cornell UP, 2001.

———. *Wasted Lives: Modernity and Its Outcasts*. Malden: Wiley, 2004.

Bennett, Jane. *Vibrant Matter: A Political Ecology of Things*. Durham and London: Duke UP, 2010.

Berthin, Christine. *Gothic Hauntings: Melancholy Crypts and Textual Ghosts.* London: Palgrave-Macmillan, 2010.

Beville, Maria. *Gothic-Postmodernism: Voicing the Terrors of Postmodernity.* Amsterdam and New York: Rodopi, 2009.

Bhandari, N. R. *25 Years of Bhopal Gas Tragedy: Inside Story and Untold Truths.* Bhopal: Lalwani, 2010.

Bhopal Express. Feature. Dir. Mahesh Mathai. 1999.

Bhopal Gas Disaster Research Centre, Medico Legal Institute, Mahatma Gandhi Medical College, Bhopal and Institute of Pathology, Indian Council of Medical Research. *Health Effects of the Toxic Gas Leak from Union Carbide Methyl Isocyanate Plant in Bhopal: Technical Report on Pathology and Toxicology (1984–1992).* New Delhi: Indian Council of Medical Research, 2010.

Bhopal Medical Appeal and Bhopal Group for Information and Action. *The Bhopal Marathon.* http://bhopal.org/wp-content/uploads/2015/03/Bhopal-Marathon-email.pdf. 20 August 2016.

Bhopal: A Prayer for Rain. Feature. Dir. Ravi Kumar. Revolver Entertainment, 2014.

Bhopal: The Search for Justice. Documentary. Dir. Peter Raymont and Lindalee Tracey. 2004.

Bhopali. Documentary. Dir. Max Carlson. 2011.

Bloom, Clive. "Horror Fiction: In Search of a Definition," in David Punter (ed.) *A New Companion to the Gothic.* Malden: Wiley-Blackwell, 2012.

Boltanski, Luc. *Distant Suffering: Morality, Media and Politics.* Cambridge: Cambridge UP, 1999.

Botting, Fred. "Gothic Culture," in Catherine Spooner and Emma McEvoy (eds.) *The Routledge Companion to the Gothic.* London and New York: Routledge, 2007. 199–213.

———. "Post-millennial monsters: Monstrosity-no-more," in Glennis Byron and Dale Townshend (eds.) *The Gothic World.* Oxford: Routledge, 2014. 498–509.

Bracke, Sarah. "Bouncing Back: Vulnerability and Resistance in Times of Resistance," in Judith Butler et al. (eds.) *Vulnerability in Resistance.* Durham and London: Duke UP, 2016. 52–75.

Brantlinger, Patrick. *Rule of Darkness: British Literature and Imperialism 1830–1914.* Ithaca: Cornell UP, 1988.

Brewster, Scott. "Rites of Defilement: Abjection and the Body Politic in Northern Irish Poetry," *Irish University Review* 35.2 (2005): 304–319.

Brink, Cornelia. "Secular icons: Looking at photographs from Nazi concentration camps," *History & Memory,* 12.1 (2000): 135–150.

Brown, Mick. "Death Trap," Letters from the Field, *The Bhopal Medical Appeal,* 2009. http://bhopal.org/stories-news/letters-from-the-field/. 20 Aug. 2016.

Browning, Jackson B. "Union Carbide: Disaster at Bhopal," http://www.bhopal.com/Reports-Studies-UCC-Opinions-and-Court-Decisions. 19 September 2016.

Butler, Judith, Zeynep Gambetti and Leticia Sabsay. "Introduction," in Butler, Gambetti and Sabsay (eds.) *Vulnerability in Resistance.* Durham and London: Duke UP, 2016. 1–11.

Butler, Judith. *Precarious Life: The Powers of Mourning and Violence.* London: Verso, 2004.

———. *Frames of War: When Is Life Grievable*? London: Verso, 2009.

———. *Notes towards a Performative Theory of Assembly.* Cambridge, MA: Harvard UP, 2015.

———. "Rethinking Vulnerability and Resistance," in Butler, Gambetti and Sabsay (eds) *Vulnerability in Resistance.* Durham and London: Duke UP, 2016. 12–27.

Byron, Glennis. "Gothic in the 1890s," in David Punter (ed.) *A New Companion to the Gothic.* Malden: Wiley-Blackwell, 2012.

Carroll, Noël. *The Philosophy of Horror, or, Paradoxes of the Heart.* London and New York: Routledge, 1990.

Castricano, Jodey. *Cryptomimesis: The Gothic and Jacques Derrida's Ghost Writing.* Montreal: McGill-Queen's UP, 2001.

Cavarero, Adriana. *Horrorism: Naming Contemporary Violence.* Trans. William McCuaig. New York: Columbia UP, 2011.

Chemical Valley. Documentary. Dir. Mini Pickering and Anne Lewis Johnson. 1991.

Chishti, Anees. *Dateline Bhopal: A Newsman's Diary of the Gas Disaster*. New Delhi: Concept Publishing, 1986.

Chouhan, T. R., et al. *Bhopal: The Inside Story. Carbide Workers Speak Out on the World's Worst Industrial Disaster*. New York and Goa: The Apex Press and The Other India Press, 1994.

Christie, Nils. "The Ideal Victim," in Ezzat A. Fattah (ed.) *From Crime Policy to Victim Policy: Reorienting the Justice System*. London: Palgrave Macmillan, 1986. 17–30.

Chute, Hillary. *Disaster Drawn: Visual Witness, Comics, and Documentary Form*. Cambridge, MA: Belknap-Harvard UP, 2016.

Cixous, Hélène. "Fiction and Its Fantoms: A Reading of Freud's *Das Unheimliche* (The 'Uncanny')," *New Literary History* 7 (1976): 525–48.

Cloud Over Bhopal. Documentary. Dir. Gerardo Olivares and Larry Levene. 2001.

Coole, Diane and Samantha Frost. "Introducing the New Materialisms," in Coole and Frost (eds.) *New Materialisms: Ontology, Agency and Politics*. Durham and London: Duke UP, 2010. 1–45.

Couser, Thomas. *Vulnerable Subjects: Ethics and Life Writing*. Ithaca: Cornell UP, 2004.

Cyr, Rachel E. "Testifying Absence in the Era of Forensic Testimony," *International Journal of Politics, Culture and Society* 26 (2013): 93–106.

D'Silva, Themistocles. *The Black Box of Bhopal: A Closer Look at the World's Deadliest Industrial Disaster*. Trafford, 2007.

Debrix, François. "Horror beyond Death: Geopolitics and the Pulverization of the Human," *New Formations* 89–90 (2017): 85–100.

Derrida, Jacques. *Spectres of Marx: The State of Debt, the Work of Mourning and the New International*. Trans. Peggy Kamuf. London: Routledge, 1994.

———. "Faith and Knowledge," in Jacques Derrida and Gianni Vattimo (eds.) *Religion*. Stanford, CA: Stanford UP, 1998. 1–78.

———. "Demeure: Fiction and Testimony," in Maurice Blanchot (ed.), *The Instant of My Death/Demeure: Fiction and Testimony*. Trans. Elizabeth Rottenberg. Stanford: Stanford UP, 13–103.

———. "Autoimmunity: Real and Symbolic Suicide," in Giovanna Borradori (ed) *Philosophy in a Time of Terror*. Chicago: U of Chicago P, 2003. 85–136.

———. *Rogues*. Trans. Pascale-Anne Brault and Michael. Stanford, CA: Stanford UP, 2005.

Derwin, Susan. "What Nazi Crimes against Humanity Can Tell Us about Torture Today," in Julie Carson and Elisabeth Weber (eds.) *Speaking About Torture*. New York. Fordham University Press, 2012.

Diederich, Nicole A. "The Gothic as Semiotic Disruption: Layers and Levels of Terror and the Abject in Mary Wilkins Freeman's 'The Wind in the Rose-bush,'" *Journal of the Midwest Modern Language Association* 44.2 (2011): 21–41.

Domanska, Eva. "Toward the Archaeontology of the Dead Body," trans. Magdalena Zapędowska, *Rethinking History* 9.5 (2005): 389–413.

Dombrowsky, Wolf R. "Again and again: Is a disaster what we call a 'disaster'?" in E. L. Quarantelli (ed.) *What Is a Disaster? Perspectives on the question*. London and New York: Routledge, 1998. 13–24.

Douglas, Mary. *Purity and Danger: An Analysis of Concepts of Pollution and Taboo*. 1966. London and New York: Routledge, 2001. Quoted from pg. 54.

Edensor, Tim. *Industrial Ruins: Spaces, Aesthetics, and Materiality*. Oxford: Berg, 2005.

Eliot, T. S. *The Waste Land*. New York: Horace Liveright, 1922.

Ewalt, Joshua P. "Mapping Injustice: *The World Is Witness*, Place-Framing, and the Politics of Viewing on Google Earth," *Communication, Culture & Critique* 4 (2011): 333–54.

Fahey, Tracey. "Blood Sugar: Gothic Bodies and Diabetes," *Gothic Studies* 17.1 (2015): 92–103.

Farnell, Gary. "The Gothic and the Thing," *Gothic Studies* 11.1 (2009) 113–23 .

Fehrenbach, Heidi, and David Rodogno (eds.). *Humanitarian Photography: A History*. Cambridge: Cambridge UP, 2015.

Freud, Sigmund. "The 'Uncanny'" in *Collected Papers*. Trans. Joan Riviere. Vol. 4. London: The Hogarth Press and the Institute of Psychoanalysis, 1971.

Generation X of the Bhopal Gas Victims: A Continuing Tragedy. Documentary. Dir. Betwa Sharma. No date.

Gilbert, Claude. "Studying disaster: Changes in the main conceptual tools," in E. L. Quarantelli (ed.) *What Is a Disaster? Perspectives on the Question*. London and New York: Routledge, 1998. 3–12.

Gitelman, Lisa. *Paper Knowledge: Towards a Media History of Documents*. Durham and London: Duke UP, 2014.

Greenpeace Research Laboratories, University of Exeter. *The Bhopal legacy: Toxic contaminants at the former Union Carbide factory site, Bhopal, India: 15 years after the Bhopal accident*. Technical Note 04/99. Labunska, I, A. Stephenson, K. Brigden, R. Stringer, R. Santillo. 1999.

Gupta, Suchandana. "Chief secretary told me to let Anderson go: Ex-collector," *The Times of India*, 10 June 2010. http://timesofindia.indiatimes.com/india/Chief-secy-told-me-to-let-Anderson-go-Ex-collector/articleshow/6029917.cms. 15 Dec. 2016.

Haider, Amna. "War Trauma and Gothic Landscapes of Dispossession and Dislocation in Pat Barker's *Regeneration* Trilogy," *Gothic Studies* 14.2: 55–73.

Hammami, Rema. "Precarious Politics: The Activism of 'Bodies That Count' (Aligning with Those That Don't) in Palestine's Colonial Frontier," in Butler, Gambetti and Sabsay (eds.) *Vulnerability in Resistance*. Durham and London: Duke UP, 2016. 167–190.

Hanna, Bridget, Ward Morehouse, Satinath Sarangi. *The Bhopal Reader: Remembering Twenty Years of the World's Worst Industrial Disaster*. Goa: Other India Press and New York: The Apex Press, 2005.

Hauck, Christina. "Abortion and the Individual Talent," *ELH* 70.1 (2003): 223–266.

Heath, Deborah, et al. "Genetic Citizenship," in David Nugent and Joan Vincent (eds.) *A Companion to the Anthropology of Politics*. Malden, MA: Blackwell, 2004. 152–67.

Hirsch, Marianne, and Leo Spitzer. "Incongruous Images: 'Before, During, and After' the Holocaust," *History and Theory* 48 (2009): 9–25.

Hirsch, Marianne. "Family Pictures: Maus, Mourning, and Post-Memory," *Discourse* 15.2 (1992–93): 3–29.

———. "Introduction: Familial Looking," in Hirsch, *The Familial Gaze*. Hanover: U Press of New England, 1999.

Hogle, Jerold E. *The Cambridge Companion to Gothic Fiction*. Cambridge: Cambridge UP, 2002.

———. "The Gothic Ghost of the Counterfeit and the Progress of Abjection," in David Punter (ed.) *A New Companion to the Gothic*. Malden: Wiley-Blackwell, 2012.

Hopkins, Lisa. *Screening the Gothic*. U of Texas P, 2005.

Horner, Avril, and Sue Zlosnik. "Comic Gothic," in David Punter (ed.) *A New Companion to the Gothic*. Malden: Wiley-Blackwell, 2012. 321–334.

Hughes, William. "'An Angel Satyr Walks These Hills': Imperial Fantasies for a Post-Colonial World," *Gothic Studies* 5.1 (2003): 121–28.

Hurley, Kelly. "Abject and Grotesque," in Catherine Spooner and Emma McEvoy (eds.) *The Routledge Companion to the Gothic*. London and New York: Routledge, 2007. 137–146.

———. *The Gothic Body: Sexuality, Materialism, and Degeneration at the* fin de siècle. Cambridge: Cambridge UP, 1994.

Huyssen, Andreas. "Nostalgia for Ruins," *Grey Room* 23 (2006): 6–21.

Jaisingh, Indira. "Legal Let-Down," in Chouhan, *Bhopal: The Inside Story*. New York and Goa: The Apex Press and The Other India Press, 1994. 181–212.

Jarenski, Shelly. "Invisibility Embraced: The Abject as a Site of Agency in Ellison's *Invisible Man*," *MELUS: Multi-Ethnic Literature of the U.S.* 35.4 (2010): 85–109.

Jenkins, Jennifer L. "'Lovelier the Second Time Around': Divorce, Desire, and Gothic Domesticity in *Invasion of the Body Snatchers*," *Journal of Popular Culture* 45.3 (2012): 478–496.

Johnson, Peter. "The Geographies of Heterotopia," *Geography Compass* 7.11 (2013): 790–803.

Jones, Christopher, Shi-Lin Loh, Kyoko Satō. "Narrating Fukushima: Scales of a Nuclear Meltdown," *East Asian Science, Technology and Society* 7.4 (2013): 601–623.

Jones, Timothy G. "The Canniness of the Gothic: Genre as Practice," *Gothic Studies* 11.1 (2009): 124–133.

Kalaidjian, Walter. "Incoming: Globalization, Disaster, Poetics," *South Atlantic Quarterly* 106.4 (2007): 825–848.

Kalelkar, Ashok S., and Arthur D. Little. "Investigation of Large-Magnitude Incidents: Bhopal as a Case Study," Institution of Chemical Engineers Conference on Preventing Major Chemical Accidents, 1988. http://www.bhopal.com/Reports-Studies-UCC-Opinions-and-Court-Decisions. 19 September 2016.

Keetley, Dawn. "Stillborn: The Entropic Gothic of *American Horror Story*," *Gothic Studies* 15.2 (2013): 89–107.

Kemp, Martin. *Christ to Coke: How Image Becomes Icon*. Oxford: Oxford UP, 2012.

Kenny, Steven. "The Uncanny Body: Blood, Flesh and the Anatomical Illustration," *Journal of Visual Communication in Medicine* 35.3 (2012): 155–157.

Kermode, Frank. *The Genesis of Secrecy*. Cambridge, MA: Harvard UP, 1979.

Khair, Tabish. *The Gothic, Postcolonialism and Otherness: Ghosts from Elsewhere*. London: Palgrave-Macmillan, 2009.

Kontsevaia, Diana B. "Mass Graves and the Politics of Reconciliation: Construction of Memorial Sites after the Srebrenica Massacre," *Totem* 21.1 (2013): 15–31.

Kristeva, Julia. *Powers of Horror: An Essay on Abjection*. Trans. and ed. Leon S. Roudiez. New York: Columbia UP, 1984.

Lapierre, Dominique, and Javier Moro. *It Was Five Past Midnight in Bhopal*. Trans. Kathryn Spink. New Delhi: Full Circle, 2001.

Latour, Bruno. *Politics of Nature: How to Bring the Sciences into Democracy*. Trans. Catherine Porter. Cambridge, MA: Harvard UP, 2004.

Laura, Sarah Juliet and Karen Embry. "A Zombie Manifesto: The Nonhuman Condition in the Era of Advanced Capitalism," *Boundary 2* 35.1 (2008): 85–108.

Linfield, Susie. *The Cruel Radiance: Photography and Political Violence*. Chicago: U of Chicago P, 2010.

Linke, Uli. "Racializing Cities, Naturalizing Space: The Seductive Appeal of Iconicities of Dispossession," *Antipode* 46.5 (2013): 1222–1239.

Linnemann, Travis, Tyler Wall, and Edward Green. "The Walking Dead and Killing State: Zombification and the Normalization of Police Violence," *Theoretical Criminology* 18.4 (2014): 506–527.

Litigating Disaster. Documentary. Ilan Ziv. 2004.

Lury, Celia. *Branding: The Logos of Global Economy*. London and New York: Routledge, 2004.

———. "Bringing the World into the World: The Material Semiotics of Contemporary Culture," *Distinktion*, 13.3 (2012): 247–260.

MacIntyre, Alasdair. *Dependent Rational Animals: Why Human Beings Need the Virtues*. 1999. Chicago: Open Court, 2002.

Malladi, Amulya. *A Breath of Fresh Air*. New York: Ballantine, 2003.

Mbembe, Achille. "Necropolitics." Trans. Libby Meintjes. *Public Culture* 15.1 (2003): 11–40.

McDonald, Brian J. "'Nothing you can believe is not coming true': Don DeLillo's *Underworld* and the End of the Cold War Gothic," *Gothic Studies* 10.2 (2008): 94–109.

McEvoy, Emma. "Gothic and the Romantics," in Catherine Spooner and Emma McEvoy (eds.) *The Routledge Companion to Gothic*. London and New York: Routledge, 2007. 19–28.

McGuire, Ann, and David Buchbinder. "The Forensic Gothic: Knowledge, the Supernatural, and the Psychic Detective," *Canadian Review of American Studies* 40.3 (2010): 289–307.

McLaughlin, Becky. "Gothicizing Apotemnophilia: Live Burial, Secret Desire, and the Uncanny Body of the Amputee Wannabe," *Word and Text* 3.2 (2013): 133–147.

Mighall, Robert. "Gothic Cities," in Catherine Spooner and Emma McEvoy (eds.) *The Routledge Companion to Gothic*. London and New York: Routledge, 2007. 54–62.

Miles, Robert. "Eighteenth-Century Gothic," in Catherine Spooner and Emma McEvoy (eds.) *The Routledge Companion to Gothic*. London and New York: Routledge, 2007. 10–18.

Millington, Nate. "Post-Industrial Imaginaries: Nature, Representation and Ruin in Detroit, Michigan," *International Journal of Urban and Regional Research* 37.1 (2013): 279–96.

Minogue, Sally, and Andrew Palmer. "Confronting the Abject: Women and Dead Babies in Modern English Fiction," *Journal of Modern Literature* 29.3 (2006): 103–125.

Mitchell, W. J. T. "Picturing Terror: Derrida's Autoimmunity," *Critical Inquiry* 33 (2007): 277–290.

"Modi Seeks to Revive India's 'Zombie Factories,' Not Abandon Them," 13 October 2014. http://in.reuters.com/article/india-industry-revival-modi-idINKCN0I202C20141013.

Monleon, José B. "1848: The Assault on Reason," in Ken Gelder (ed.) *The Horror Reader*. London and New York: Routledge, 2000. 20–28.

Moore, Alexandra Schultheis. "'Disaster Capitalism' and Human Rights: Embodiment and Subalternity in Indra Sinha's *Animal's People*," in Elizabeth Swanson Goldberg and Alexandra Schultheis (eds.) *Theoretical Perspectives on Human Rights and* Literature. New York and London: Routledge, 2013. 231–246.

Moore, Francesca. *Bhopal: Facing 30*. London: Arts Council England, 2014.

Morgan, Jack. *The Biology of Horror: Gothic Literature and Film*. Carbondale and Edwardsville: Southern Illinois UP, 2002.

Morrison, Susan Signe. *The Literature of Waste: Material Ecopoetics and Ethical Matter*. New York: Palgrave-Macmillan, 2015.

Morton, Tim. *Hyperobjects: Philosophy and Ecology after the End of the World*. Minneapolis and London: U of Minnesota P, 2013.

Mukherjee, Rahul. "Remembering Bhopal: Situated Testimony, Chronic Disaster, and Memories of Survival," *Media Fields Journal* 5 (2012). http://mediafieldsjournal.squarespace.com/remembering-bhopal/. Accessed 15 August 2016.

Mukherjee, Suroopa. *Surviving Bhopal: Dancing Bodies, Written Texts, and Oral Testimonials of Women in the Wake of Industrial Disaster*. London: Palgrave-Macmillan, 2010.

Murphy, Ann V. "Corporeal Vulnerability and the New Humanism," *Hypatia* 26.3 (2011): 575–590.

Naas, Michael. "'One Nation . . . Indivisible': Jacques Derrida on the Autoimmunity of Democracy and the Sovereignty of God," *Research in Phenomenology* 36 (2006): 15–44.

Nayar, Pramod K. "Of Homes and the 'Unhomely': Michael Cunningham's Specimen Days," *Americana* 8.2 (2012). http://americanaejournal.hu/vol8no2/nayar.

———. "From Bhopal to Biometrics: Biological Citizenship in the Age of Globalization," in Scott Slovic, R. Swarnalatha, and Vidya Sarveswaran (eds.) *Ecoambiguity, Community and Development: Towards a Politicized Ecocriticism*. Lanham: Lexington, 2014. 85–98.

———. "Elementals: The Arts of Bhopal, 1984–2015," *eSocial Sciences* 29 November 2015.

———. "Autobiogenography: Genomes and Life Writing," *a/b: Auto/biography Studies* 31.3 (2016): 509–525.

———. "The Child Victim as the Face of Modern Suffering," *The Wire* 24 Sept. 2016. https://thewire.in/67883/meaning-is-a-physiognomy-or-the-child-victim-as-the-face-of-the-modern-age/.

———. "Dystopia, Waste and the Decadent Sublime," *Indian Journal of English Studies* LIV (2017): 32–42.

Neocleous, Mark. "Resisting Resilience," *Radical Philosophy* 178 (2013): 2–7.

Nguyen, Vinh-Kim. "Antiretroviral Globalism, Biopolitics, and Therapeutic Citizenship," in Aihwa Ong and Stephen J. Collier (eds.) *Global Assemblages: Technology, Politics, and Ethics as Anthropological Problems*. Malden, MA: Blackwell, 2005. 124–144.

Nixon, Rob. *Slow Violence and the Environmentalism of the Poor*. Cambridge, MA: Harvard UP, 2011.

Noys, Benjamin. *The Culture of Death*. Oxford: Berg, 2005.

Odysseos, Louiza. "The Question Concerning Human Rights and Human Rightlessness: Disposability and Struggle in the Bhopal Gas Disaster," *Third World Quarterly* 36.6 (2015): 1041–1059.

One Night in Bhopal. Documentary. Dir. Steve Condie. BBC, 2004.

Pal, Mahuya, and Mohan J. Dutta. "Organizing Resistance on the Internet: The Case of the International Campaign for Justice in Bhopal," *Communication, Culture & Critique* 5 (2012): 230–251.

Perera, Suvendrini. "'They Give Evidence': Bodies, Borders and the Disappeared," *Social Identities* 12.6 (2006): 637–656.

Petryna, Adriana. *Life Exposed: Biological Citizens after Chernobyl*. Princeton, NJ: Princeton UP, 2002.
Pikner, Tarmo. "Enactments of Urban Nature: Considering Industrial Ruins," *Geografiska Annaler: Series B, Human Geography* 96.1 (2014): 83–94.
Porfiriev, Boris N. "Issues in the definition and delineation of disasters and disaster areas," in E. L. Quarantelli (ed.) *What Is a Disaster? Perspectives on the Question*. London and New York: Routledge, 1998. 51–68.
Prajapati, H. L. *Gas Tragedy: An Eyewitness*. New Delhi: Mittal Publications, 2003.
Presto, Jenifer. "The Aesthetics of Disaster: Blok, Messina, and the Decadent Sublime," *Slavic Review* 70.3 (2011): 569–590.
Punter, David, and Glennis Byron. *The Gothic*. Oxford: Blackwell: 2007.
Punter, David. "Shape and Shadow: On Poetry and the Uncanny," in David Punter (ed.) *A New Companion to the Gothic*. Malden: Wiley-Blackwell, 2012. 252–264.
Rai, Raghu. *Exposure: Portrait of a Corporate Crime*. Amsterdam: Greenpeace, 2002.
Rajagopal, Arvind. "And the Poor Get Gassed: Multinational-Aided Development and the State—The Case of Bhopal," *Berkeley Journal of Sociology* 32 (1987): 129–152.
Raymond, Michael David. *John Wordsworth on Snowdon: The Elegiac Sublime and the Spectacle of Woe*. Fordham University. http://fordham.bepress.com/dissertations/AAI3353777/. Proquest.
Renshaw, Layla. "The Dead and Their Public. Memory Campaigns, Issue Networks and the Role of the Archaeologist in the Excavation of Mass Graves," *Archaeological Dialogues* 20 (2013): 35–47.
Round, Julia. "Gothic and the Graphic Novel," in David Punter (ed.) *A New Companion to the Gothic*. Malden: Wiley-Blackwell, 2012. 335–349.
Royle, Nicholas. *The Uncanny*. Manchester: Manchester UP, 2003.
Salgado, Minoli. "Vanishing Points/Visible Fictions: The Textual Politics of Terror," *Textual Practice* 27.2 (2013): 207–223.
Sargent, Lyman Tower. "The Three Faces of Utopianism Revisited," *Utopian Studies* 5.1 (1994): 1–37.
Saxon, Wolfgang. "Union Carbide Says Upset Worker Set Off the Bhopal Plant Disaster," *The New York Times* 11 August 1986. http://www.nytimes.com/1986/08/11/world/union-carbide-says-upset-worker-set-off-the-bhopal-plant-disaster.html. 18 September 2016.
Schultheis, Alexandra W. *Regenerative Fictions: Postcolonialism, Psychoanalysis, and the Nation as Family*. London: Palgrave-Macmillan, 2004.
Seconds from Disaster: Bhopal Nightmare. Documentary. Dir. Russell Eatough. National Geographic, 2011.
Sekula, Allan. "Reading the Archive: Photography between Labor and Capital," in Jessica Evans and Stuart Hall (eds.) *Visual Culture*. London and Thousand Oaks: Sage—The Open University, 1999. 181–192.
Shapira, Yael. "Where the Bodies Are Hidden: Ann Radcliffe's 'Delicate' Gothic," *Eighteenth-Century Fiction* 18.4 (2006): 1–24.
Shapiro, Stephen. "Transvaal, Transylvania *Dracula*'s World-System and Gothic Periodicity," *Gothic Studies* 10.1 (2008): 29–47.
Shastri, Lalit. *Bhopal Disaster: An Eyewitness Account*. New Delhi: Criterion, 1985.
Shukin, Nicole. *Animal Capital: Rendering Life in Biopolitical Times*. Minneapolis and London: U of Minnesota P, 2006.
———. "The Biocapital of Living—and the Art of Dying—After Fukushima," *Postmodern Culture* 26.2 (2016).
Singh, Moti. *Unfolding the Betrayal of Bhopal Gas Tragedy*. Delhi: BR Publishing, 2008.
Sinha, Indra. *Animal's People*. London: Pocket Books—Simon and Schuster, 2007.
Sloterdijk, Peter. *Terror from the Air*. Trans. Amy Patton and Steve Corcoran. Cambridge, MA: MIT P, 2009.
Smith, Alan Lloyd. "The Phantoms of *Drood* and *Rebecca*: The Uncanny Reencountered through Abraham and Torok's 'Cryptonymy,'" *Poetics Today* 13.2 (1992): 285–308.
Spencer, Kathleen L. "Purity and Danger: Dracula, the Urban Gothic, and the Late Victorian Degeneracy Crisis," *ELH* 59.1 (1992): 197–225.

Spooner, Catherine. "Goth Culture," in David Punter (ed.) *A New Companion to the Gothic*. Malden: Wiley-Blackwell, 2012. 350–366.

Srishti. *Surviving Bhopal 2002: Toxic Present—Toxic Future: A Report on Human and Environmental Chemical Contamination around the Bhopal Disaster Site*. 2002.

Star, Susan Leigh, and James R. Griesemer, "Institutional Ecology, 'Translations' and Boundary Objects: Amateurs and Professionals in Berkeley's Museum of Vertebrate Zoology, 1907–39," *Social Studies of Science* 19.3 (1989): 387–420.

Strobl, Rainer. "Becoming a Victim," in Shlomo Giora Shoham, Paul Knepper, Martin Kett (eds.) *The International Handbook of Victimology*. CRC Press, 2010. 3–26.

Sturken, Marita. "Absent Images of Memory: Remembering and Reenacting the Japanese Internment," *Positions* 5.3 (1997): 687–707.

Sundaram, Neeraja. "Imagining Bio-disaster, Reproducing Social Order: Epidemics in Contemporary Hollywood," *Journal of Creative Communications* 7.1–2 (2012): 135–151

Tagg, John. *The Disciplinary Frame: Photographic Truths and the Capture of Meaning*. Minneapolis and London: U of Minnesota P, 2009.

Talairach-Vielmas, Laurence. "'I Have Bottled Babes Unborn': The Gothic, Medical Collections and Victorian Popular Culture," *Gothic Studies* 17.1 (2015): 28–42.

Tatar, Maria M. "The Houses of Fiction: Toward a Definition of the Uncanny," *Comparative Literature* 33 (1981): 167–82.

Taylor, Diana. *Disappearing Acts: Spectacles of Gender and Nationalism in Argentina's "Dirty War."* Durham and London: Duke UP, 1997.

Tézenas, Ambroise. *I Was Here: Photographs of Dark Tourism*. Dewi Lewis, 2015.

Thapa, L. B. *Bhopal Gas Tragedy: A Survivor's Tale*. Memoir. Kindle ed. No date.

The Bhopal Chemical Disaster: Twenty Years without Justice. Documentary. Dir. Sanford Lewis. 2004.

The Bhopal Disaster. National Geographic. 2014.

Thomson, Rosemarie Garland. *Extraordinary Bodies: Figuring Physical Disability in American Culture and Literature*. New York: Columbia UP, 1997.

Tippins, D. J., M. P. Mueller, M. van Eijck, J. D. Adams (ed.). *Cultural Studies and Environmentalism: The Confluence of EcoJustice, Place-Based (Science) Education, and Indigenous Knowledge Systems*. Springer Science & Business Media, 2010.

Toogood, Colin. "Water, the Ongoing Crisis," *777* (2011): 7–9.

Townshend, Dale. "Introduction," Glennis Byron and Dale Townshend (eds.) *The Gothic World*. Oxford: Routledge, 2014. xiv–xlvi.

Turner, Bryan S. *Vulnerability and Human Rights*. University Park: University of Pennsylvania Press, 2006.

Turvey, Samuel T., and Anthony S. Cheke. "Dead as a Dodo: The Fortuitous Rise to Fame of an Extinction Icon," *Historical Biology* 20.2 (2008): 149–163.

Twomey, Christina. "Framing Atrocity: Photography and Humanitarianism," *History of Photography* 36.3 (2012): 255–264.

Tyler, Imogen. *Revolting Subjects: Social Abjection and Resistance in Neoliberal Britain*. London: Zed, 2013.

Van Dijck, José. "Medical Documentary: Conjoined Twins as a Mediated Spectacle," *Media Culture and Society* 24 (2002): 537–556.

Varma, Rahul. *Bhopal*. Toronto: Playwrights Canada P, 2004.

Verdrey, Katherine. *The Political Lives of Dead Bodies: Reburial and Postsocialist Change*. New York: Columbia UP, 1999.

Vice, Sue. *Introducing Bakhtin*. Manchester: Manchester UP, 1997.

Vidler, Anthony. *The Architectural Uncanny: Essays in the Modern Unhomely*. Cambridge, MA: MIT, 1992.

Wacquant, Loïc. "Plumbing the Social Underbelly of the Dual City," 2015. www.loicwacquant.net. 29 Dec. 2016.

———. "Revisiting Territories of Relegation: Class, Ethnicity and State in the Making of Advanced Marginality," *Urban Studies* 53.6 (2016): 1077–1088.

Wagner, Tamara S. "Clinical Gothic: Sensationalising Substance Abuse in the Victorian Home," *Gothic Studies* 11.2 (2010): 51–74.

———. “Gothic and the Victorian Home,” in Glennis Byron and Dale Townshend (eds.) *The Gothic World*. London and New York: Routledge, 2014. 110–120.

Warsop, Andrew. “The Ill Body and *das Unheimliche* (the Uncanny),” *Journal of Medicine and Philosophy* 36 (2011): 484–495.

Waskul, Dennis D., and Pamela van der Riet. “The Abject Embodiment of Cancer Patients: Dignity, Selfhood, and the Grotesque Body,” *Symbolic Interaction* 25.4 (2002): 487–513.

Wasson, Sara. “Gothic Cities and Suburbs: 1880—the Present,” in Glennis Byron and Dale Townshend (eds.) *The Gothic World*. Oxford: Routledge, 2014. 132–142.

———. “Scalpel and Metaphor: The Ceremony of Organ Harvest in Gothic Science Fiction,” *Gothic Studies* 17.1 (2015): 104–123.

Weber, Samuel. “The Sideshow, or: Remarks on a Canny Moment,” *MLN* 88 (1973): 1102–133.

———. *Targets of Opportunity: On the Militarization of Thinking*. Fordham UP, 2005.

Weisman, Steven R., and Sanjoy Hazarika. “Theory of Bhopal Sabotage Is Offered,” *New York Times* 23 June 1987. http://www.nytimes.com/1987/06/23/world/theory-of-bhopal-sabotage-is-offered.html?pagewanted=all. 17 September 2016.

Williams, Tony. *Hearths of Darkness: The Family in the American Horror Film*. Madison, NY: Farleigh Dickinson UP, 1996.

Wright, Terence. *Visual Impact: Culture and the Meaning of Images*. Oxford: Berg, 2008.

Yaeger, Patricia. “Trash as Archive, Trash as Enlightenment,” in Gay Hawkins and Steven Muecke (eds.) *Culture and Waste: The Creation and Destruction of Value*. Lanham: Rowman & Littlefield, 2003. 103–116.

Young, Craig, and Duncan Light. “Corpses, dead body politics and agency in human geography: Following the corpse of Dr. Petru Groza,” in *Transactions of the Institute of British Geographers* 38 (2013): 135–148.

Index

About the Author

Pramod K. Nayar teaches in the Department of English at the University of Hyderabad, India. His most recent books include *The British Raj: Keywords* (2017), *The Extreme in Contemporary Culture* (2017), *Human Rights and Literature* (2016), *The Indian Graphic Novel* (2016) and *The Postcolonial Studies Dictionary* (2015). His work on graphic novels and memoirs have appeared in *Biography, Image and Text, Journal of Postcolonial Writing* and other journals and anthologies. In addition to Bhopal, he has published on posthumanism, genetic autobiographies, colonial discourse, human rights narratives, cultural studies and several subjects in English literature—in journals such as *Modern Fiction Studies, a/b, South Asia, Asiatic, Ariel, Postcolonial Text, South Asian Review, 1650–1850, Celebrity Studies, Orbis Litterarum, Prose Studies, Changing English, ANQ, Notes on Contemporary Literature*, and *Explicator*.

CPSIA information can be obtained
at www.ICGtesting.com
Printed in the USA
BVOW03*0949151117
499698BV00007B/2/P